INSECT PEST MANAGEMENT: Readings

Edited by
Merle Shepard
Clemson University

MSS Information Corporation
655 Madison Avenue, New York, N.Y. 10021

This is a custom-made book of readings prepared for the courses taught by the editor, as well as for related courses and for college and university libraries. For information about our program, please write to:

MSS INFORMATION CORPORATION
655 Madison Avenue
New York, New York 10021

MSS wishes to express its appreciation to the authors of the articles in this collection for their cooperation in making their work available in this format.

Library of Congress Cataloging in Publication Data

Shepard, Merle, comp.
Insect pest management.

CONTENTS: Pest management background: Southwood, T. R. E. and Way, M. J. Ecological background to pest management. Geier, P. W. Management of insect pests. Geier, P. W. and Clark, L. R. An ecological approach to pest control. Clark, L. R. Analysis of pest situations through the life systems approach. [etc.)
1. Insect control--Addresses, essays, lectures.
I. Title.
[SB931.S492] 632'.7'08 73-9730
ISBN 0-8422-7114-7
ISBN 0-8422-0296-X (pbk.)

TABLE OF CONTENTS

PESTICIDES

SAMPLING METHODS

PEST MANAGEMENT SYSTEMS

PREFACE

The concept of insect pest management stemmed mainly from the serious inadequacies and disturbances which arose from sole reliance on chemical pesticides and the realization that many of the natural regulating agents can be manipulated to suppress harmful insect species. The literature is becoming more replete with case histories of pesticidal failures and their attendant undesirable side effects. The shifting of emphasis from *control* of pest insects to the more ecologically-based idea of *management* implies an understanding of the forces in nature which reduce populations of noxious species to tolerable levels. The preponderance of past research efforts in the applied area of entomology has been oriented toward immediate reduction or elimination of insect pest population, i.e., chemical control. The transition to a more encompassing alternative has now evolved.

The study of insect pest management must be approached from a standpoint of long range planning. Immediate elimination of chemical pesticides will not solve the insect pest problems. Their proper use in an integrated management system will not only have a more stabilizing influence on our man-made monocultures but will also allow researchers time to develop alternative methods with more permanent, less disruptive, environmental influence.

Basic to the philosophy of pest management are methods of censusing populations of destructive species and their biotic regulating agents and relating these population estimates to damage by the former. Development of sequential sampling models and establishment of economic thresholds may provide a sound and reliable foundation from which more refined and sophisticated computer simulation models may evolve.

I have made an attempt to avoid articles dealing with the theoretical aspects of insect pest management, although population biology is basic to the understanding of management concepts. This collection of readings by no means covers the spectrum of management practices but is only an attempt to acquaint the student of insect pest management with some of the tools available for application and implementation of a management program.

M.S.

Clemson, South Carolina
June, 1973

PEST MANAGEMENT BACKGROUND

ECOLOGICAL BACKGROUND TO PEST MANAGEMENT

by

T. R. E. SOUTHWOOD AND M. J. WAY

The basis of the management of a pest is the planned manipulation of the various processes that influence the economic injury level, so as to minimise the economic effect of the pest. The economic damage inflicted by a pest is compounded of the value of a crop and its yield, the latter is the product of the interaction of the pest population with the plant. Different varieties of plants and the same variety under different growing conditions can vary in their tolerance of the 'eating rate' of the same size population of a pest; this is an aspect of plant resistance and is one of the interactions that can be manipulated in pest management.

All the other processes are concerned with the size of the pest population. In essence the population system of any animal will depend on the action of the five components: climate, food, space, other animals of different sorts, and the population itself and other populations of the same species acting through the pathways of natality, mortality and dispersal (Andrewartha & Birch 1954; Southwood 1969a). However, a population system does not function in isolation, its components are expressions of the ecosystem of which it is a part. Man-made agricultural habitats or agroecosystems have certain peculiar features brought about by cultivation.

In this paper the special feature of agroecosystems are discussed, then the various population processes that might be manipulated reviewed. It is shown that the potentiality for pest management of the different processes varies according to certain characters, notably diversity, stability and isolation, that occur to markedly differing degrees in different types of agroecosystems. These special characters are analysed and the role of the type of agroecosystem in influencing the choice of management processes indicated.

SPECIAL FEATURES OF AGROECOSYSTEMS

There are a great range of natural ecosystems, from the diverse, permanent rain forest to the simple, permanent arctic tundra or to the diverse, but unstable (temporary) desert ephemeral community. Equally, man-made ecosystems vary in regard to permanence, isolation and diversity; nevertheless in respect of certain features they generally differ from natural ecosystems.

Continuity

Agroecosystems are not self-perpetuating and so they are always of limited duration; they vary from orchards that may last half a century to

CONCEPTS OF PEST MANAGEMENT, Proceedings of Conference, March 25-27, 1970, pp. 6-28.

arable crops that may be harvested a few months after planting. They have an abrupt start in ground that has been cleared of other vegetation, usually by ploughing or burning. Natural ecosystems are self-perpetuating; some are virtually permanent, others like desert ephemerals or secondary scrub are more temporary, but often continuous in space.

Selection of Vegetation

In natural ecosystems the vegetation is the product of natural selection so it is that best adapted to withstand the various climatic and edaphic factors; the individual plants are exposed to intense competition, especially when young. In crops, the plants are selected by man and when young are usually surrounded by much bare ground so that competition between them is minimal (e.g. in acid grassland in England there are commonly 40 million stems per acre, but in a cereal crop the average number of stems is 1.5 million per acre). Agroecosystems can suffer catastrophically from climatic variations (e.g. the effect of late frosts on citrus in Florida and on apples in England).

Diversity of Species

Man normally plants a single species, occasionally two (inter-cropping), very rarely several (mixed orchards); modern herbicides often eliminate weeds and thus agroecosystems are usually monocultures. Natural ecosystems mostly consist of several species of plant, although exceptionally, particularly in extreme climates, relatively pure stands of one or two species may occur (e.g. coniferous forests, papyrus swamps).

Intraspecific Diversity (Age and Varieties)

The planting practices for most crops ensure that in agroecosystems the vegetation is of a uniform age; likewise the varieties planted have usually been selected so that there is a very limited range of genotype, and thus the synchronisation of flowering, leaf-burst and other processes is much more marked than in a natural ecosystem.

Nutrient and Water Supply

Many crops are fertilized with chemicals or manure and some are irrigated. The levels of nutrients in their foliage and amount of young growing tissue is usually greater than in natural ecosystems.

Pest and Disease Outbreaks

These are a regular feature of agroecosystems. They occur occasionally in secondary tropical forests (Conway & Tay 1969), unstable arid ecosystems (Uvarov 1957) and simple coniferous forests (Blais 1965), but are unrecorded from climax tropical rain forests (Voute 1946). This is clearly of great relevance to pest management and it will be discussed further later.

THE POPULATION SYSTEM

Starting with the colonization of a plant by a phytophagous insect and following the development of the population throughout subsequent seasons, the sequence of population process is:

1. Invasion of the crop.
2. The survival and development rates of the 'invaders' or 'residents' prior to reproduction.
3. The birth rate and survival during reproduction.
4. The survival and development rates of the immature stages.
5. The numbers emigrating.

The relationships of these various processes and some environmental variables can be seen in Fig. 1 & 2 and the factors that influence the processes in Table 1. A similar diagram would provide a sub-model for the natural enemies. A consideration of the diagrams shows immediately why pest management is such a valuable approach — the numbers of the pests are the product of a series of complex and inter-related factors, the system contains lags and feed-backs.

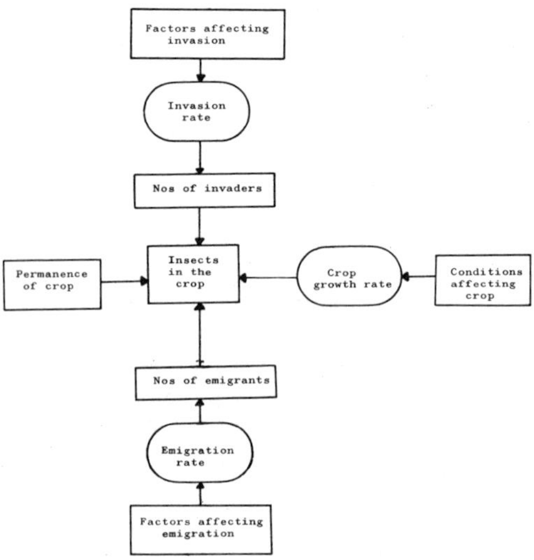

1

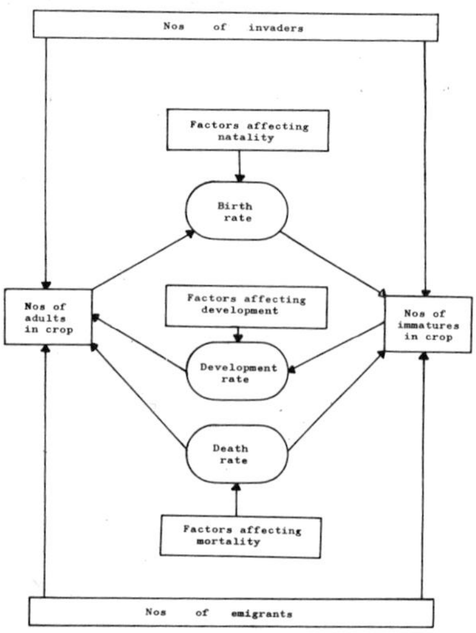

2

Fig. 1. Factors that impinge on the pest population in the crop.

Fig. 2. The population system of the pest.

Table 1. Factors affecting various population processes (Figs. 1 & 2)

Factors affecting invasion

Numbers of potential invaders
Distance of source of invaders
Conditions for invasion and settling
Attractiveness of crop

Factors affecting natality, mortality, development and emigration

Food, quality and quantity
Climate
Pesticides
Animals of the same sort (intra-specific competition, sex-ratios, virility)
Organisms of other sorts (natural enemies, interspecific competitors)

The manipulation of this population system in pest management will have two aspects:

1. The prevention of rapid population increase, i.e., outbreaks.
2. The lowering of the general level of the population.

The characters of the population processes that are important for each of these aspects are different. Outbreaks can only be controlled by directly density dependent (or regulating) processes that will destroy an increasing proportion of the population as density rises: parasites, predators, and competitive effects may be directly density dependent. It is inherent in the properties of pesticides that they kill a fixed proportion of the population and so are not in themselves able to regulate a pest. If they are to even approximately regulate a pest then the applications must be limited to periods of population increase. The "scouting method" is a management procedure that assists in this approach (Mathews & Tunstall 1968). Outbreaks may arise from the effect of disturbing (or destabilising) factors; in pest management the aim is to lessen the role of these.

The level of the population may be reduced by:

1. Increasing the severity of a factor that acts equally at all densities (e.g. increasing a kill of 30% to 60%).
2. Varying the conditioning factors that determine the framework within which the density dependent (or regulating) factors operate (Huffaker & Messenger 1964; Richards & Southwood 1968).
3. Reducing the number of animals that arrive in the crop and limiting their development rate and the time available to them for reproduction.

The extent to which each approach is adopted and the level that can be tolerated will depend on the damage potential per individual pest. A virus carrier or other low density pest has a high potential, but a leaf miner, for example, will normally have a low potential and a relatively high level of population can be tolerated.

One of the important properties of population systems, from the point of view of pest management, is that when the overall level is lowered, regulating factors that were previously unable to control the population may be able to do so (Takahashi 1964). Thus pest management is concerned with every type of population process: regulating, disturbing, and conditioning.

FACTORS IN A POPULATION SYSTEM OF PARTICULAR SIGNIFICANCE IN AGROECOSYSTEMS

Number of Potential Invaders

The numbers of a pest that are available to invade a newly planted crop, will depend on the size of its population and its level of migratory movement. Insects that live on stable (permanent) vegetation have a lower level of migratory movement, more specific feeding habits and, often, lower potential rates of increase than those from unstable (temporary) habitats (Southwood 1962). Conway (1970) has noted in Sabah that pests from the latter type of habitat invade a new crop much more readily than those that live on the former.

The level of migratory activity in a population may suddenly rise if its habitat changes. Douglas (1939) has suggested that invasions of arable crops by pests follows the destruction of wild vegetation; the sudden invasion of cotton by *Dysdercus* and *Calidea* follows changes in their alternate hosts (Pearson 1958). The cutting and harvesting of alfalfa leads to massive migrations of the spittlebug, *Philaenus* (Wiegert 1964) and the weevil, *Hypera* (Pamanes and Pienkowski 1965; Prokopy and Gyrisco 1965), movements that do not occur if the crop remains uncut and succulent.

The actual relationship between the number of potential invaders and the magnitude of the infestation on the crop has been investigated for *Aphis fabae;* outbreaks on field beans could be predicted by the numbers overwintering on certain spindle trees, *Euonymus europaeus*, in the vicinity (Way & Banks 1968). In contrast, on a larger scale, Strickland (1960) calculated that the numbers of beet leafhoppers in S. Idaho and E. Oregon and the numbers of frit fly in England and Wales each spring, were in the range of ten to a hundred-fold of those necessary to infest all crop lands. In temperate regions by the end of the growing season the number of pests produced from the crop lands can be greatly in excess of those from natural habitats as was shown in the frit fly by Southwood & Jepson (1962).

Distance from Source of Invaders

In general the dilution principle holds so that crops near sources of invaders are likely to receive heavier infestations than those at greater distance. However, there are considerable differences, between species, in the amount of horizontal displacement typical for an individual migrant (Johnson 1969). The average distance travelled each day by a frit fly, *Oscinella frit*, in England in August has been calculated as 20 miles; this is an insect that flies up into the air and is borne on air currents at 1,000 ft. or more (Johnson et al 1962; Southwood et al 1961). In central U.S.A. aphids probably travel several hundreds of miles (Taylor 1965). In contrast other insects generally keep below their boundary layer and travel much smaller distances, e.g., wheat bulb fly, *Leptohylemyia coarctata* (Bardner et al 1968) and cotton boll weevil (Fye et al 1959), where most individuals appear to move relatively few miles. Hence, excluding the effects of weather and topography, the distance needed to isolate a crop from a source of invaders can vary from a few to over a thousand miles. The same principles apply to natural enemies as to pests.

Conditions for Invasion and Settling

The extent to which a population of potential invaders takes-off and the distance they travel will be influenced by current weather conditions. As Johnson (1969) points out, in his comprehensive review, there are considerable variations in response from species to species, but cold and rain generally reduce take-off. Many pests are borne on the prevailing winds and therefore the direction of the wind in relation to the source of the invaders will determine the numbers above a given habitat; for example, Brown (1965) has suggested that this is the mechanism that determines which areas in the Middle East are regularly, occasionally, or seldom, infested by cereal bugs. The actual accumulation and deposition of the airborne insects is influenced by weather fronts and by immediate shelter and edge effects (Johnson 1969); many insects are deposited on the leeward side of wind breaks, fences and hedges (Lewis 1965, 1966, 1967).

Attractiveness of the Crop

When insects are below their boundary layer, i.e. in virtually complete control of their displacement, they are able to respond to various stimuli by settling. The stimuli may be visual (colour, shape, colour contrast) or olfactory. For example certain aphids are particularly attracted to yellow

(Moericke 1950; Kennedy & Stroyan 1959; Cartier 1965, 1966), flower dwelling thrips to white (Lewis 1959), and cabbage butterflies to green and blue-green (Ilse 1937). The contrast provided by a vertical cereal shoot against its background influences the frit fly, *Oscinella frit*, and indeed it seems as if this contrast principle is widespread for in many insects it has been shown that the infestation level per plant (and often per area) is lower on dense plantings than sparse ones [e.g. *Cephus* (Luginbill & McNeal 1958), cabbage pests (Pimentel 1961), *O. frit* (Southwood unpublished)]. Extensive studies on aphids and whiteflies on Brussels sprouts grown against different backgrounds (bare soil, weeds, artificial) have confirmed that the practice of growing them 'strikingly silhouetted against the bare soil is seemingly the best way to attract colonising aphids' (Smith 1969 a & b). There are fewer examples of the role of olfactory stimuli even over short distances in plant feeders [*Ostrinia* (Schurr and Holdaway 1965), *Dysdercus* (Saxena 1965)]. More extensive evidence has been obtained with various wood-boring beetles (Wood 1962; Norris 1965; Vité and Gara 1962). Work on the mechanism of host plant selection by insects has been reviewed by Thorsteinson (1960), Beck (1965), Kennedy (1965) and Schoonhoven (1968).

Components of the sequence of response to stimuli leading to the choice of prey and oviposition site by a natural enemy are in principle similar to those when the prey (or host) selects the host plant. There is, however, the additional stage associated with the selection of the prey. Many of the components of the stimulus pattern leading to oviposition by natural enemies have been elucidated by work on aphid predators, notably by Blackman (1967), Chandler (1968a, b, c & d), Iperti (1965) and Smith (1969a & b). Whereas some predators are attracted to crop plants growing in bare soil, others only appear in numbers when there are many weeds in the crop. Predators differ also in their response to the prey, when selecting oviposition sites: some species of syrphid will lay freely on uninfested plants, others prefer plants with large aggregations.

The series of properties that may serve as stimuli in the selection of habitat, host plant, prey and oviposition site are shown in Table 2.

Food Supply, Quality and Quantity

The plant may actually cause the death of a phytophagous insect by biophysical or biochemical means: this is an aspect of plant resistance to insect attack and was termed by Painter (1951) "antibiosis". Hooked hairs (Johnson 1953), sticky exudates and resin (Harris 1960), rapidly proliferating wound tissue (Oatman 1959) and a range of complex biochemicals (Beck 1965) are among the mechanisms. Although such resistance can be enhanced by plant breeding, it can also be lost "accidently" in programmes designed to enhance other characters (van Emden 1966a).

Crops are irrigated and treated with chemical fertilisers and various growth regulators: these operations will change the constituents of the plants and so affect the rate of development and fecundity of insects living on them. Although there is considerable variation from species to species and with the different constituents, the changes in quality brought about by cultivation generally favour the pest (Strickland 1960; van Emden 1966b).

Perhaps one of the most important features in the crop habitat favouring the development of outbreaks of phytophagous species is the continuity of large numbers of plants of the same species, for this enables excess individuals produced on one plant to find a new host plant with the minimum of exposure to the hazards of dispersal.

Table 2. Some of the properties that may serve as stimuli in the selection of habitat, host plant, prey, and oviposition site by phytophagous insects and their natural enemies.

Selection of:		Property
Habitat	Optical	(1) Colour of and colour contrasts within stand.
		(2) Size and shape of plant stand.
		(3) Density of stand.
	Olfactory	(4) Odour of host plant stand.
Host plant or host plant of prey or host.	Optical	(5) Colour of and colour contrast of host plant with general habitat.
		(6) Size and shape of host plant.
	Olfactory	(7) Odour of host plant.
	Gustatory	(8) Taste of host plant (rarely relevant to natural enemy).
	Tactile	(9) Physical properties of surface of plant.
Prey or host.	Optical	(10) Colour of and colour contrast of prey (host) with plant background.
		(11) Shape and size of prey (host).
		(12) Movement of prey (host).
	Olfactory	(13) Odour of prey (host).
	Gustatory	(14) "Taste" of prey (host) (only where adult feeds on prey (host)).
Ovi-position site	Optical	(15) Shape of plant parts or of host.
		(16) Movement of prey (host).
	Tactile	(17) Actual site of oviposition on plant or in host.
	Gustatory	(18) Composition of plant or host detected by ovipositor.

Competition, Intra- and Inter-specific

There is no evidence that the conditions in agroecosystems have any particular effect on intraspecific competition, although it can be important in phytophagous insects, even at moderate densities (Way 1968; Way & Banks 1967). The same applies to interspecific effects: there are examples of one phytophagous species competing with another so as to virtually replace it [e.g. the occurrence of *Bemisia* on cotton when *Empoasca* populations are reduced (Ripper & George 1965)]. It is therefore theoretically possible that a form of biological control could be developed in which a species that caused damage to the crop was replaced by one that was of less economic significance (Beirne 1962; Clausen 1966; McGovran et al 1969).

Natural Enemies

The effects of predators, parasites and disease can be very important in a pest management programme. Firstly we will consider the theoretical relationship between the role of the natural enemy and the numbers of its prey. If a natural enemy acts in a density dependent manner, that is, kills an increasing proportion of the prey as density increases, it is considered to be a regulating (or stabilising) process, unless it overcompensates (Southwood 1966). When this stabilising response occurs within a single generation, the predator is said to be directly density dependent and to show a behavioural response to the change in prey density (Hassell 1966). Such a response, by a species of natural enemy, can arise in two ways: by each individual killing more of the prey, a 'functional' or 'individual' response or by an increase in the numbers of the 'enemy' attacking the prey, an 'aggregative' or 'numerical' response (Solomon 1949; Holling 1959; Hassell 1966). As Murdoch (1970) has pointed out, the individual response may be behavioural or may be due to increased consumption with increased growth and development. The density related response of a natural enemy can however arise in the next or subsequent generations, through the effects of increased prey density on its reproduction; this is referred to as a delayed density dependent process (Varley 1953) and is one aspect of the numerical response as originally defined by Solomon (1949). If a natural enemy kills a smaller proportion of the prey population as the latter increases, it is termed an inversely density dependent process and will have the effect of increasing any population fluctuations.

A particular natural enemy will not always respond in the same way; the response can vary with habitat and the level of the prey population. An example of the former is *Trichogramma* that appears to have a direct response to the density of the host, *Diatraea*, in some areas, but not in others (Southwood 1969b). Field studies, for example Clark (1964) on psyllids and Readshaw (1965) on phasmatids, show that phytophagous pests can be said to have two different population levels: the endemic or lower level where various natural enemies may normally regulate the population and the epidemic or higher level where intra-specific effects ultimately bring about population decline. Takahashi (1964) has shown theoretically how this arises by the prey escaping from regulation; its rate of increase suddenly increases, normally due to some change in the environment (the disturbing factor) and it outstrips the ability of the natural enemy to respond. As prey population density soars, the predator will cease to be a direct density dependent, stabilising factor and may become inversely density dependent and unstabilising in its action.

Such then are the basic mechanisms of a predator-prey system; its action in an agricultural habitat will be influenced by a whole series of factors similar to those already reviewed for the pest. They will be briefly reconsidered in this new context.

TIME AVAILABLE FOR INVASION. Often natural enemies do not establish themselves in the newly planted crop as quickly as their prey. There is therefore an initial period when the pest has arrived on the crop, but is virtually free from natural enemies; the longer the crop is in existence the greater the chance of the natural enemies becoming established. *Dasynus piperis* is a pest in new pepper gardens, but after some time it is controlled by hymenopterous parasites (van der Vecht 1933), likewise the cyclamen mite, *Steneotarsonemus pallidus*, causes damage in young strawberry fields, whereas in older fields it is effectively controlled by the predatory mite, *Typhlodromus* (Huffaker & Kennett 1956).

SOURCES OF INVADERS. Soil cultivation causes heavy mortality of most plant feeding insects and of the pupal stages of natural enemies (Way et al 1969) ; newly cultivated fields are therefore devoid, with few exceptions, of non-soil insects. The principles of invasion and deposition have already been discussed. (p. 9) : a nearby source of natural enemies will greatly enhance their role in the crop as was shown by Doutt & Nakata (1965) for the mymarid parasite of the grape leafhopper in California.

FOOD AVAILABLE FOR NON-ENTOMOPHAGOUS STAGES. The adults of many dipterous and hymenopterous parasites require pollen and nectar to mature their eggs or to prolong life and increase fecundity (Leius 1960) and hence nearby prey populations may be attacked more heavily (van Emden 1965 a & b).

FOOD FOR ENTOMOPHAGOUS STAGES. Alternate prey can serve as a source of natural enemies (Hambleton 1944; Doutt & Nakata 1965) and as a food source when the pest density is low (Taylor 1937).

INTERSPECIFIC COMPETITION. The number of different predators and parasites preying on a pest is yet another aspect of diversity, that will be discussed in general below. There is indeed a controversy in the biological control field between those who believe the introduction of a single species is best (Turnbull & Chant 1961; Zwolfer 1963), and those who support multiple introductions (Huffaker & Kennett 1966). It is probably impossible to generalise (Doutt & DeBach 1964), for as Hassell & Varley (1969) have shown the outcome will depend on the specific properties of the various parasites.

INTRA-SPECIFIC COMPETITION. A review of this complex topic is quite outside the scope of this paper: the subject has however recently been greatly clarified by Hassell & Varley (1969) who have shown that the equation for the relationship between the parasite population density and that of its host must include a term "the mutual interference constant".

DIVERSITY AND STABILITY

Throughout the above discussion, reference has frequently been made to other host plants, to other phytophagous species and to the number of species of natural enemies — that is, to the diversity of the ecosystem. Initially it was pointed out that tropical rain forests are very diverse in terms of plant and animal species and the numbers of animals within them do not fluctuate violently. Starting from this observation Elton (1958) and others have suggested that stability is a consequence of diversity and thus the reversal of the present trend towards simplification in agriculture is essential. In contrast Watt (1964, 1968) analysed the records of the Canadian Forest Insect Survey and showed that the more competitors a herbivore had, the more stable its population, but the more host trees the less stable its population. Watt extended these conclusions to the predator-prey system and implied that they conflict with the "traditional wisdom of ecologists".

We consider that for any ecosystem the greater the diversity the greater the chances of stability, in other words, that violent fluctuations will not occur in the numbers of any species within it. It is impossible to generalise further, diversity is a broad term and covers a wide range of relationships and these must be distinguished.

In the first place, the distinction must be made between interspecific diversity and trophic diversity (Southwood 1966, p. 350; Watt 1968, p. 45). Diversity at the interspecific level (i.e. at one trophic level) is a straightforward concept, the more species, the more diverse. But it does not formally convey any information about the interrelations of these species. Trophic diversity is based on the number of links between different trophic levels.

Reconsidering Watts' (1964) data: this shows and much observation supports the view, that at the primary consumer level (P C) a system with inter-specific competition, as shown in Fig. 3A will be more stable than one without (Fig. 3B). This conclusion has then been extended to the next two trophic levels i.e. the primary consumer (herbivore) and secondary consumer (carnivore), which may not be justified. Furthermore, Watt (1968, p. 43) implies that the intensity of competition between the species at the higher trophic level, responsible for their stability, will 'tend to hold down pressure' against the lower level and so allow this level to fluctuate. It could be that the observed stability at the higher level is a reflection of the control this level exercises on the lower. Hassell & Varley (1969) argue that diversity at the parasite level is a consequence of stabilising mechanisms.

The second situation that Watt (1964) studied was for polyphagous species, whose numbers he found tended to fluctuate. This system (3C) is exactly the opposite to 3A and to describe this as diverse is to extend the term to trophic interactions, moving against the flow of energy, in a way that is confusing.

We could introduce a further condition to the definition of trophic diversity: account must be taken of the direction of energy flow. Only if there are more species (more 'energy outlets') at the higher level is the system more diverse. Thus 3D is more diverse than 3E. There is one further variable whose effect on diversity we must define, it is the number of cross-links. Comparing Fig. 3D and 3F, at any level the inter-specific diversity between the primary and secondary consumer levels is greater in 3F than in 3D.

However, the stability of a system will depend not only on its trophic diversity, but on the actual nature of the trophic links, on the extent to which the species at the higher level can respond to increases in the lower level. In other words on the precision and extent of the density dependence in the link.

This theoretical treatment can be illustrated by a few specific examples. Locusts and grasshoppers are unstable species, not because of their large range of natural enemies, but primarily because they live in unstable habitats (Uvarov 1957; Southwood 1962). They are also polyphagous, so even on an *a priori* consideration of their trophic system (Fig. 3G) they would not be expected to be very stable. A classical example of biological control which indicates that factors other than mere numbers of species at different trophic levels are important is the biological control of the leaf-mining hispid beetle, *Promecotheca reichei* in Fiji. This species was con-

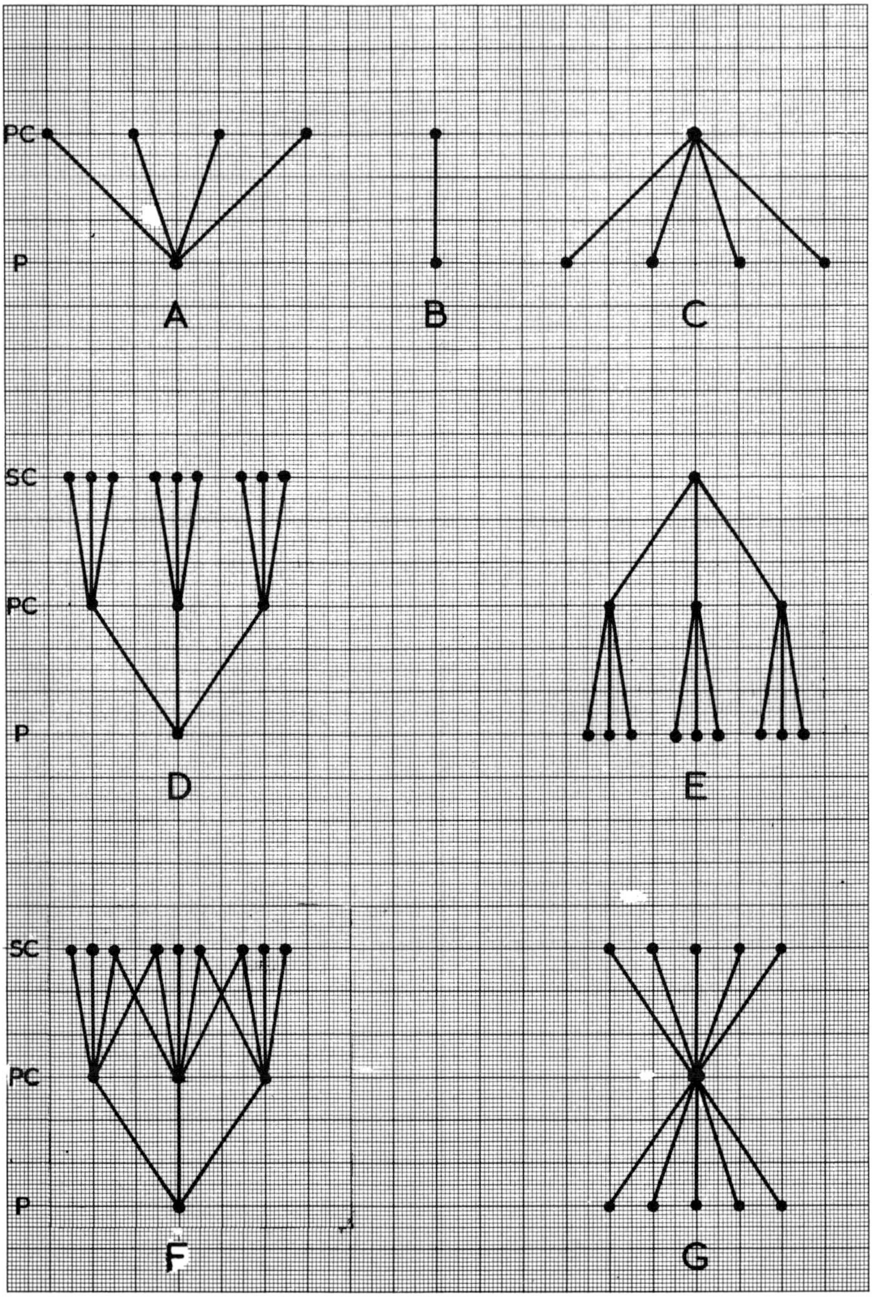

Fig. 3. Schematic representation of the trophic systems of differing diversity (see text for discussion).

trolled at a stable low density by a complex of indigenous natural enemies until a predaceous mite was accidently added to the complex. This new species upset the continuity of the prey's generation cycle, upon which the other natural enemies depended. They became rare and the beetle became a serious pest. Nevertheless the addition of a further natural enemy species to the entomophagous trophic level re-established stability and within a year the beetle returned to stable scarcity (Taylor 1937). Another example of a phytophagous species that can be maintained at low and stable populations by a complex of predators is the fruit-tree red-spider mite *(Paranychus ulmi)*, for which Collyer (1953) records forty-five species of predator. Biological control also can be successful when the predator is monophagous (e.g. *Rodolia cardinalis* on *Icerya purchasi*) and when the pest is without competitors [e.g. spider mites and *Myzus persicae* in greenhouses (Hussey 1965)].

In conclusion trophic diversity, as defined here, is likely to promote stability, but whether it does so will depend on the climatic stability of the habitat and the precision of the response of any particular trophic link to an increase in the population at the lower level.

ISOLATION AND STABILITY

The dispersion of the host plants, their degree of isolation, is another factor that affects stability (Watt 1968, p. 47). It interacts with diversity through the effect of the permanence of the habitat on the time available for invasion. It is necessary to distinguish between the isolation of the crop and the isolation of individual plants, both can, in extreme cases, promote population stability but for differing reasons.

If a crop is well isolated, by distance and in respect to the 'pest-bearing-winds' (p. 11), from other crops or sources of invaders, it may escape pests altogether — stability due to nil population. The chances of invasion will increase with time, and therefore this management method is generally only practical with annual crops in areas that have an inclement season. This approach is the normal practice for the production of virus-free seed potatoes in Scotland (Broadbent 1967).

In contrast, in very permanent ecosystems, such as tropical rain forests, the isolation of each individual tree — the fragmentation of the 'crop' leading to a mosaic — ensures that each population has only a small habitat that it can occupy before becoming crowded. These circumstances accentuate dispersal mortality at a relatively low overall population. In the extreme case, if the animal only occurs in very small numbers per plant (i.e. a low density pest) and plants are well separated, then 'dispersal losses' could virtually exterminate the population.

CHOICE OF MANAGEMENT PROCEDURE IN RELATION TO THE CHARACTERS OF THE AGROECOSYSTEM

Pest management involves the integration of biological, cultural and chemical methods of control in an integrated approach (Stern et al 1959; Smith & van den Bosch 1967). The importance of the various factors in the population system already reviewed (Fig. 1 & 2, Tables 1 & 2) and the potentialities for the different methods of control depend in general on four characteristics of the agroecosystem:

1. The diversity of vegetation within the agroecosystem.
2. The permanence of the various crops within the agroecosystem.
3. The stability of the climate.

4. The extent of the isolation of the agroecosystem from natural vegetation.

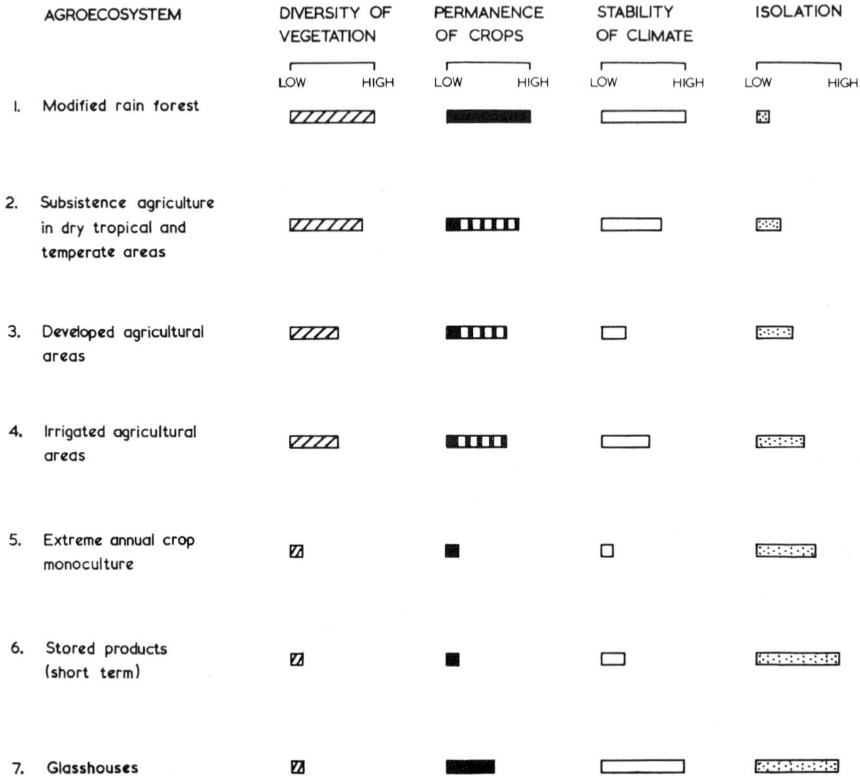

Fig. 4. The characteristics, of particular significance in pest management programmes, of the main types of agroecosystems.

A series of seven types of agroecosystem are compared in respect of these characters in Fig. 4. The following generalisations are possible:

1. The need and opportunities for manipulation as opposed to mimicking nature, increase through the series of agroecosystems 1-7.
2. The danger of upsetting natural biological control (by autochthonous natural enemies), through the use of pesticides decreases through the series of agroecosystems from 1-5 & 6.
3. The opportunities for manipulated biological control in the isolated agroecosystems (4-7) are highest in greenhouses (7), moderate in irrigated areas (4) and lowest in monocultures (5) and stored products (6). They are directly correlated with the stability of the climate.
4. The role of "planting date" as a management procedure is greatest in those agroecosystems towards the bottom of the figure (i.e. 3-5), for here the problems of invasion are greatest (because of degree

of isolation), and the crops can be managed so that they pass through the susceptible stage simultaneously. This is a particularly powerful approach when the plant is only susceptible to or suffers damage from the insect for a brief period. For example young oats are susceptible to the frit fly only when in the 2-4 leaf stage and many arboreal species really flourish only on young foliage (Simanton 1965; White 1967). It should be noted that the even-aged and genetically uniform crop does provide exceptionally favourable conditions, if these should synchronise with the appropriate stage of the pest. Massive synchronisation between pest and crop in any year is an important cause of outbreaks and may be more significant than variations in the absolute population of potential invaders. Where the timing of the invasion is determined by cultural practices in other crops (e.g. spittlebugs; Wiegert, 1964) the practices in the two crops should be co-ordinated to ensure the minimum of synchronisation.

Diversity

The theoretical discussion above (p. 14) of the role of ecosystem diversity in population stability emphasised the impossibility of firm prediction in the absence of detailed knowledge of the density response of the various trophic links. In the more practical context, any management practice affecting diversity must be assessed separately in respect of its effects on the pest and those on its natural enemies. The full diversity and stability of climax vegetation does promote stability at the consumer level, particularly in a stable and favourable climate, but in most agroecosystems this diversity has been reduced to a greater or lesser extent. In these simplified conditions it is not justified to assume that any increase in diversity will have a stabilising effect (Way 1966; Kennedy 1968). Thus the diversity conferred by hedgerows in agricultural areas may favour pests more than natural enemies or *vice versa* (van Emden 1965a). Some major pests in the U.K., such as the black bean aphid *(Aphis fabae)* would be greatly reduced, if not eliminated, should their alternate host plants be removed from hedgerows and other wild areas. The availability of pollen and nectar from flowers in these areas greatly influences the fecundity of some vegetable pests [e.g. the cabbage root fly *Erioischia brassicae* (Finch & Coaker 1969)]. and it is of similar importance to a much larger number of natural enemies (van Emden 1965a). Certain cultural practices, such as the regular cutting of verges and removal of bottom flora from hedgerows seem to affect predatory species more than the phytophagous ones (Southwood & van Emden 1967; Pollard 1968a & b).

The aims of pest management should be to determine what elements of diversity need to be retained or added and what need to be eliminated to enhance the management of the pests' population. Doutt and Nakata (1965) describe how the simple addition of a small area of a single species of blackberry into Californian vineyards provided all the diversity needed to ensure biological control of a leaf-hopper. More extensive intermingling of other species is necessary for the biological control of the coconut pest, *Pseudotheraptus wayi*, in Zanzibar: other species of tree need to be planted between the coconuts to provide better nesting conditions for the predatory ant, *Oecophylla*, and an alternate source of food for it, particularly honeydew from coccids (Way 1953, 1954, 1963). *P. wayi* is a low density pest — a population of less than 1.5 per palm can cause complete loss of crop; such a pest can normally be controlled biologically only by a polyphagous predator with other dependable foods (in this case other insects and honeydew).

The Permanence of the Crops

It will be appreciated from the profound differences in qualities of perennial and annual crop conditions, that equally profound differences in approaches to pest management will be required. Many major pests of perennial crops have a limited tendency for dispersal and form relatively closed populations together with their complex of natural enemies (Southwood 1962; Conway 1970). The comparative stability of the plant habitat makes this possible and creates a situation where biological control by natural enemies can have an important stabilising effect on pest species. Even in places (e.g. temperate orchards) where dormancy is induced by a cold winter, the close proximity of overwintering sites of pests and natural enemies minimises the delay in response by the latter. In contrast in annual crops of limited duration, it is the level of natality (largely fixed by the success of the invasion), not that of mortality, that determines population size (Southwood 1967).

There are many examples of successful biological control in more permanent agroecosystems ranging from the classical ones, such as the control of *Icerya purchasi* on citrus and of several pests on coconuts in Fiji (Taylor 1935, 1937; Tothill, Taylor & Paine 1930), to recent ones with semi-perennial herbaceous crops such as alfalfa (Schlinger & Dietrick 1960) and strawberries (Huffaker & Kennett 1953). In the latter example the establishment of stability following planting of a crop is accelerated by artificially introducing the pest and parasite into the strawberry bed. A similar form of stability can be introduced into greenhouses where a succession of crops in the same greenhouse creates 'perennial' conditions (Hussey 1965) and sometimes in the field: in Peru, for example, the interplanting of cotton and maize in irrigated conditions in an otherwise equable climate creates a stable and self contained environment for long enough to effect biological control of the cotton boll worm, *Heliothis virescens*, and of other pests (Wille 1958). A complex situation exists in Africa where, in the equable climate of southern Uganda, *Heliothis armigera* breeds throughout the year on a wide range of crops and wild plants: the complex of small native plots in a semi-wild environment seemingly simulates 'perennial' stability ensuring that *H. armigera* remains a minor pest (Coaker 1960). Further south in Tanzania, however, the dry season induces diapause, the effect of which is to produce an "annual crop", which no doubt prohibits successful natural biological control (Reed 1965). In these circumstances the planting of maize with the cotton increases the abundance of *H. armigera* because the pest multiplies on maize and then migrates to cotton without being checked by natural enemies which in Peru multiply during the host's generations on maize. Thus the same crop practices minimise the effects of the annual crop instability in some circumstances, but exaggerate it in others.

It is significant that in suitable climatic and crop situations, control by natural enemies is possible with *H. virescens* and *H. armigera* because they are polyphagous on a succession of annual plant hosts thereby mimicking the continuity provided by a perennial plant host.

In contrast, species like the pink bollworm are specific to cotton, which is normally grown seasonally. In these circumstances natural enemy action appears to be insignificant, not only because there are no alternate hosts for the bollworm, but also because the chosen control measures involve the enforcement of a definitive 'closed season' for cotton. This management procedure aims to make conditions directly intolerable for the pest itself by increasing the instability of the environment rather than relying on the indirect reduction of its population through natural enemies.

Another example of the value of cultural practices in enhancing the instability of a crop and so aiding in the management of a species with very few wild hosts is provided by the tobacco hornworms (*Manduca* spp.). The practice of eliminating sucker growth by applications of maleic hydrazide during the period of crop maturation also greatly reduced post-harvest sucker growth. This comprised a drastic limitation of food during the period when diapausing individuals are produced and markedly reduced the population level (Rabb 1969).

With brassica pests, their ecology is contrastingly different in two geographical areas: a situation that rather parallels that in cotton. In parts of South Africa with an equable climate, where the crops and their associated insects are actively developing throughout the year, Ulyett (1947) pointed to the need for growing successive brassica crops close together to maintain the stability for effective control by natural enemies. In Britain however, the winter season enforces a complete break in the development of the natural enemies, and soil cultivations between September and April destroy virtually all overwintering parasites and predators in the field (Way et al 1969). The process of invasion has to recommence next season and thus it is not surprising that pest damage is worst in areas where a succession of brassica crops is concentrated.

Climate

Variations in climate are probably the commonest disturbing factors in population systems (Varley 1963; Southwood 1967; Richards & Southwood 1968). Therefore population stability is most likely in equable and favourable climates, where the number of incipient outbreaks (due to disturbing factors) to which the natural enemies will have to respond, is minimal. Readshaw's (1965) theory of phasmatid outbreaks in the Australian Eucalyptus forests is a good example of the disturbing role of climate. An exceptionally cool summer in an area adversely affects the egg parasites and causes a local increase in stick-insects. The many species of birds, that are predators, move into this area of abundant prey — an aggregative response (p. 13). This reduces predation in the surrounding regions, allowing the phasmid population there to rise above its normal level; these areas then attract further predators from more outlying regions. The outbreak appears to spread out like shock waves from an epicentre. The initial disturbance was climatic in origin, but its spread is due to the disturbance of the predator/prey system. Because the phasmatid has a two year life cycle (and hence outbreaks occur every second year in outbreak areas) the predators are prevented from responding numerically through increased overall population.

In general, man, at present, can do little to influence climate except in greenhouses and with elaborate irrigation systems. Bytinski-Salz (1965) records that in Israel increased overhead irrigation of crops can aid the establishment of the crawler stage of coccids, but is detrimental to their hymenopterous natural enemies and so causes outbreaks of scale insects. In less arid conditions, in other countries, the humidity can be raised to a higher level which is optimal for entomophagous fungi which then control the scale insects.

Isolation

The measure of isolation of a crop can seriously affect the natural enemy component of an integrated control programme. Such programmes may be successful in temperate regions where in spite of the disturbance due to the cold or dry season, the availability of hibernation sites and/or alternate prey nearby allows a rapid reinvasion of the crop by natural enemies. Integrated control programmes for apple pests appear to have been successful in Nova Scotia (Pickett & Patterson 1953; MacPhee & Sandford 1956) and in Virginia (Clancy & McAlister 1956). Similar programmes failed in the Okonagon Valley in British Columbia (Morgan & Anderson 1957) and this may well have been in part due to this valley being a small irrigated area surrounded by arid scrub and desert, so there were no "refuges" for the natural enemies. (Increased fungal disease of the apples was the factor to which Morgan and Anderson attributed importance).

If natural enemies are to be used in pest management programmes in isolated crops, especially in irrigated areas, then the appropriate diversity must be introduced into the agroecosystem itself. This has been done in California by the practice of strip-cropping in alfalfa (Schlinger & Dietrick 1960). Indeed this emphasises the general point that diversity needs to be managed.

With low density pests (e.g. the coconut bug, *Pseudotheraptus wayi*, where 1.5 per tree constitutes an outbreak), any particular habitat has to be uniformly favourable over a relatively large area to support a viable population; in other words the minimum size of a population territory is large and excessive 'fragmentation' will lead to heavy dispersal mortality. Little damage from low density pests is therefore suffered by isolated plants or by plants that are surrounded by others that are protected. Thus with *P. wayi* if 70% or more of the palms in a plantation are protected by ants, the rest of the palms escape serious attack (Way 1953, 1954).

Acknowledgements

We are grateful to our colleagues, especially Dr. G. R. Conway, for valuable discussions on the general substance of this review and helpful criticism of the graphical presentations.

References

Andrewartha, H. G. and Birch, L. C. 1954. The Distribution and Abundance of Animals. Univ. of Chicago Press.

Bardner, R., Lofty, J. R., Huston, P. and Maskell, F. E. 1968. Movement and oviposition of wheat bulb flies. Pl. Path. 17, 97-103.

Beck, S. D. 1965. Resistance of plants to insects. Ann. Rev. Ent. 10, 207-232.

Beirne, B. P. 1962. Integration of biological and chemical control. Desirable attributes of biotic agents. Bull. ent. Soc. Amer. 8, 189-191.

Blackman, R. L. 1967. Selection of aphid prey by *Adalia bipunctata* L. and *Coccinella 7 – punctata* L. Ann. appl. Biol. 59, 331-338.

Blais, J. R. 1965. Spruce budworm outbreaks in the past three centuries in the Laurentide Park, Quebec. Forest Sci. 11, 130-138.

Broadbent, L. 1967. Non chemical methods of disease control. *In* "The Future of Crop Protection". Proc. 2nd ann. Confr. Univ. Reading agric. Club. 51-60.

Brown, E. S. 1965. Notes on the migration and direction of flight of *Eurygaster* and *Aelia* species (Hemiptera, Pentatomoidea) and their possible bearing on invasions of cereal crops. J. Anim. Ecol. 34, 93-107.

Bytinski-Salz, H. 1965. Effects of modern agrotechnical methods on the agricultural insect pest populations in Israel. Proc. XII int. Congr. Ent. (London 1964), 583.

Cartier, J. J. 1965. Action des radiations arangées sur le puceron du pois, *Acyrthosiphon pisum* (Harr.) et sur le puceron de la pomme de terre *Macrosiphum euphorbiae* (Thos.) en élevage sur diète synthétique. Phytoprotection 46, 65-73.

Cartier, J. J. 1966. Aphid responses to colors in artificial rearings. Bull. ent. Soc. Amer. 12, 378-380.

Chandler, A. E. F. 1968a. Some host plant factors affecting oviposition by aphidophagous Syrphidae (Diptera). Ann. appl. Biol. 61, 415-423.

Chandler, A. E. F. 1968b. The relationships between aphid infestations and oviposition by aphidophagous Syrphidae (Diptera). Ann. appl. Biol. 61. 425-434.

Chandler, A. E. F. 1968c. Some factors influencing the occurrence and site of oviposition by aphidophagous Syrphidae (Diptera). Ann. appl. Biol. 61, 435-446.

Chandler, A. E. F. 1968d. Height preferences for oviposition of aphidophagous Syrphidae (Diptera). Entomophaga 13, 187-195.

Clancy, D. W. and McAlister, H. J. 1956. Selective pesticides as aids to biological control of apple pests. J. econ. Ent. 49, 196-202.

Clark, L. R. 1964. The population dynamics of *Cardiaspina albitextura* (Psyllidae). Aust. J. Zool. 12, 362-380.

Clausen, C. P. 1966. The competitive displacement and coexistence principles. Ann. Rev. Ent. 11, 183-212.

Coaker, T. H. 1960. Investigations on *Heliothis armigera* (Hb.) in Uganda. Research Memoir no. 36, Empire Cotton Growing Corporation, London.

Collyer, E. 1953. Biology of some predatory insects and mites associated with the fruit tree red spider mite (*Metatetranychus ulmi* (Koch)) in South-eastern England. II. Some important predators of the mite. J. hort. Sci. 28, 85-97.

Conway, G. R. 1970. Ecological aspects of pest control in Malaysia. *In* Farvar, M. T. and Milton, J. P. (eds.). The Careless Technology: Ecology and International Development. Natural History Press.

Conway, G. R. and Tay, E. B. 1969. Crop pests in Sabah, Malaysia and their control. State Ministry of Agriculture and Fisheries Sabah, Malaysia.

Douglas, W. A. 1939. Studies of rice stink bug populations with special reference to local migration. J. econ. Ent. 32, 302-303.

Doutt, R. L. and DeBach, P. 1964. Some biological control concepts and questions. *In* DeBach, P. (ed.). Biological Control of Insect Pests and Weeds. 118-142. Chapman and Hall, London.

Doutt, R. L. and Nakata, J. 1965. Parasites for control of grape leafhopper. Calif. Agric. 19, 3.

Elton, C. S. 1958. The Ecology of Invasions by Animals and Plants. Methuen, London.

van Emden, H. F. 1965a. The role of uncultivated land in the biology of crop pests and beneficial insects. Scient. Hort. 17, 121-136.

van Emden, H. F. 1965b. The effect of uncultivated land on the distribution of cabbage aphid *(Brevicoryne brassicae)* on an adjacent crop. J. appl. Ecol. 2, 171-196.

van Emden, H. F. 1966a. Plant resistance to insects induced by environment. Scient. Hort. 18, 92-102.

van Emden, H. F. 1966b. Plant insect relationships and pest control. Wld. Rev. Pest Control 5, 115-123.

Finch, S. and Coaker, T. H. 1969. Comparison of the nutritive values of carbohydrates and related compounds to *Erioischia brassicae.* Ent. exp. & appl. 12, 441-453.

Fye, R. E., McMillan, W. W., Walker, R. L. and Hopkins, A. R. 1959. The distance into woods along a cotton field at which the boll weevil hibernates. J. econ. Ent. 52, 310-312.

Hambleton, E. J. 1944. *Heliothis virescens* as a pest of cotton, with notes on host plants in Peru. J. econ. Ent. 37, 660-666.

Harris, P. 1960. Production of pine resin and its effect on survival of *Rhyacionia buoliana* (Schiff.) (Lepidoptera, Olethreutidae). Can. J. Zool. 38, 121-130

Hassell, M. P. 1966. Evaluation of parasite or predator response. J. anim. Ecol. 35, 65-75.

Hassell, M. P. and Varley, G. C. 1969. New inductive population model for insect parasites and its bearing on biological control. Nature, Lond. 223, 1133-1137.

Holling, C. S. 1959. The components of predation as revealed by a study of small mammal predation on the European pine sawfly. Can. Ent. 91, 293-320.

Huffaker, C. B. and Kennett, C. E. 1953. Developments towards biological control of cyclamen mite on strawberries in California. J. econ. Ent. 46, 802-812.

Huffaker, C. B. and Kennett, C. E. 1956. Experimental studies in predation: predation and cyclamen mite populations on strawberries in California. Hilgardia 26, 191-222.

Huffaker, C. B. and Kennett, C. E. 1966. Studies of two parasites of olive scale *Paralatoria oleae* (Colvee). IV Biological control of *Parlatoria oleae* (Colvée) through the compensatory action of two introduced parasites. Hilgardia 37, 283-335.

Huffaker, C. B. and Messenger, P. S. 1964. Chapters 3 and 4 *in* DeBach, P. (ed.). Biological Control of Insect Pests and Weeds. Chapman and Hall, London.

Hussey, N. W. 1965. Possibilities for integrated control of some glasshouse pests. Ann. appl. Biol. 56, 347-350.

Ilse, D. 1937. New observations on responses to colours in egg laying butterflies. Nature, Lond. 140, 544-545.

Iperti, G. 1965. Contribution a l'etude de la specificité chez les principales Coccinelles aphidiphages des Alpes — Maritimes et de Basses — Alpes. Entomophaga 10, 159-178.

Johnson, B. 1953. The injurious effects of the hooked epidermal hairs of French beans (*Phaseolus vulgaris* L.) on *Aphis craccivora* Koch. Bull. ent. Res. 44, 779-788.

Johnson, C. G. 1969. Migration and Dispersal of Insects by Flight. Methuen, London.

Johnson, C. G., Taylor, L. R. and Southwood, T. R. E. 1962. High altitude migration of *Oscinella frit* L. (Diptera: Chloropidae). J. Anim. Ecol. 31, 373-383.

Kennedy, J. S. 1965. Mechanisms of host plant selection. Ann. appl. Biol. 56, 317-322.

Kennedy, J. S. 1968. The motivation of integrated control. J. appl. Ecol. 5, 492-499.

Kennedy, J. S. and Stroyan, H. L. G. 1959. Biology of aphids. Ann. Rev. Ent. 4, 139-160.

Leius, K. 1960. Attractiveness of different foods and flowers to the adults of some hymenopterous parasites. Can. Ent. 92, 369-376.

Lewis, T. 1959. The annual cycle of *Limothrips cerealium* Haliday (Thysanoptera) and its distribution in a wheat field. Ent. exp. & appl. 2, 187-203.

Lewis, T. 1965. The effects of an artificial windbreak on the aerial distribution of flying insects. Ann. appl. Biol. 55, 503-512.

Lewis, T. 1966. An analysis of components of wind affecting the accumulation of flying insects near artificial windbreaks. Ann. appl. Biol. 58, 365-370.

Lewis, T. 1967. The horizontal and vertical distribution of flying insects near artificial windbreaks. Ann. appl. Biol. 60, 23-31.

Luginbill, P., Jr. and McNeal, F. H. 1958. Influence of seedling density and row spacings on the resistance of spring wheats to the wheat stem sawfly. J. econ. Ent. 51, 804-808.

MacPhee, A. W. and Sandford, K. H. 1956. The influence of spray programs on the fauna of apple orchards in Nova Scotia. X Supplement to VII. Effects on some beneficial arthropods. Can. Ent. 88, 632-634.

Mathews, G. A. and Tunstall, J. P. 1968. Scouting for pests and the timing of spray applications. Cott. Gr. Rev. 45, 115-127.

McGovran, E. R. et al. 1969. Insect-Pest Management and Control. Princiciples of Plant and Animal Pest Control. 3. (Nat. Acad. Sci. Pub. 1695) Washington.

Moericke, V. 1950. Über den Farbensinn der Pfirsichblattlaus *Myzodes persicae* Sulz. Z. Tierpsychol. 7, 265-274.

Morgan, C. V. G. and Anderson, N. H. 1957. Some aspects of a ryania-glyodin spray schedule in British Columbia apple orchards. I. Entomological, horticultural and economic aspects. Can. J. Pl. Sc. 37, 423-433.

Murdoch, W. W. 1970. The development response of predators to changes in prey density. (personal communication).

Norris, D. M. 1965. In-flight dispersal and orientation of two *Scolytus* species (Coleoptera) to their host plants for oviposition purposes. Proc. XII int. Congr. Ent. (London 1964), 293.

Oatman, E. R. 1959. Host range studies of the melon leaf miner, *Liriomyza pictella* (Thomson) (Diptera: Agromyzidae). Ann. ent. Soc. Amer. 52, 739-741.

Painter, R. H. 1951. Insect Resistance in Crop Plants. Macmillan. New York.

Pamanes, A. G. and Pienkowski, R. L. 1965. Dispersal of the alfalfa weevil in Virginia. Ann. ent. Soc. Amer. 58, 230-233.

Pearson, E. G. 1958. The Insect Pests of Cotton in Tropical Agriculture. Commonwealth Institute of Entomology, London.

Pickett, A. D. and Patterson, N. A. 1953. The influence of spray programs on the fauna of apple orchards in Nova Scotia. IV. A review. Can. Ent. 85, 472-478.

Pimentel, D. 1961. The influence of plant spatial patterns on insect populations. Ann. ent. Soc. Amer. 54, 61-69.

Pollard, E. 1968a. Hedges. II. The effect of removal of the bottom flora of a hawthorn hedgerow on the fauna of the hawthorn. J. appl. Ecol. 5, 109-123.

Pollard, E. 1968b. Hedges. III. The effect of removal of the bottom flora of a hawthorn hedgerow on the Carabidae of the hedge bottom. J. appl. Ecol. 5, 125-139.

Prokopy, R. J. and Gyrisco, G. G. 1965. Diel flight activity of migrating alfalfa weevils, *Hypera postica* (Coleoptera: Curculionidae). Ann. ent. Soc. Amer. 58, 642-647.

Rabb, R. L. 1969. Environmental manipulations as influencing populations of tobacco hornworms. Proc. Tall Timbers Conference on Animal Control by Habitat Management No. 1, 175-191. (Tall Timbers Research Station, Tallahassee, Florida).

Readshaw, J. L. 1965. A theory of phasmatid outbreak release. Aust. J. Zool. 13, 475-490.

Reed, W. 1965. *Heliothis armigera* (Hb.) (Noctuidae) in western Tanganyika. II. Ecology and natural and chemical control. Bul. ent. Res. 56. 127-140.

Richards, O. W. and Southwood, T. R. E. 1968. The abundance of insects: Introduction *in* Southwood, T. R. E. (ed.) Insect Abundance. R. ent. Soc. Lond. Symp. 4, 1-7.

Ripper, W. E. and George, L. 1965. Cotton pests of the Sudan, their habits and control. Blackwell Scientific Publications, Oxford.

Saxena, K. N. 1965. Control of the orientation and feeding behaviour of red cotton bug, *Dysdercus koenigii* (F.) by chemical constituents of plants. Proc. XII int. Congr. Ent. (London 1964)). 294.

Schlinger, E. I. and Dietrick, E. J. 1960. Biological control of insect pests aided by strip farming alfalfa in experimental program. Calif. Agric. 14 (1), 8.

Schoonhoven, L. M. 1968. Chemosensory bases of host plant selection. Ann. Rev. Ent. 13, 115-136.

Schurr, K. M. and Holdaway, F. G. 1965. The relation between plant injury in maize and ovipositional preference of *Ostrinia nubilalis* (Hbn.) (European Corn Borer). Proc. XII int. Congr. Ent. (London 1964) 607-608.

Simanton, W. A. 1965. Mite populations in Florida citrus groves before and after severe freezes. Proc. XII int. Congr. Ent. (London 1964), 364-366.

Smith, J. G. 1969a. Effects of crop background on populations of Brassica aphids and their natural enemies. Ph.D. Thesis University of London.

Smith, J. G. 1969b. Some effects of crop background on populations of aphids and their natural enemies on brussels sprouts. Ann. appl. Biol. 63, 326-330.

Smith, R. F. and van den Bosch, R. 1967. Integrated control. *In* Kilgore, W. W. and Doutt, R. L. (ed.). Pest Control. Biological, physical and selected chemical methods. Academic Press, New York and London.

Solomon, M. E. 1949. The natural control of animal populations. J. Anim. Ecol. 18, 1-35.

Southwood, T. R. E. 1962. Migration of terrestrial arthropods in relation to habitat. Biol. Rev. 37, 171-214.

Southwood, T. R. E. 1966. Ecological Methods. Methuen, London.

Southwood, T. R. E. 1967. The interpretation of population change. J. Anim. Ecol. 36, 519-529.

Southwood, T. R. E. 1969a. The abundance of animals. Inaug. Lect. Imp. Coll. Sci. Technol. VIII 1-16.

Southwood, T. R. E. 1969b. Population studies of insects attacking sugar cane. *In* Williams, J. R. et al (eds). Pests of Sugar Cane. Elsevier Publishing Co., Amsterdam.

Southwood T. R. E., Jepson, W. F. and van Emden, H. F. 1961. Studies on the behaviour of *Oscinella frit* L. (Diptera) adults of the panicle generation. Ent. exp. and appl. 4, 196-210.

Southwood, T. R. E. and Jepson, W. F. 1962. The productivity of grasslands in England for *Oscinella frit* L. (Chloropidae) and other stem boring Diptera. Bull. ent. Res. 53, 395-407.

Southwood, T. R. E. and van Emden, H. F. 1967. A comparison of the fauna of cut and uncut grasslands. Z. angew. Ent. 60, 188-198.

Stern, V. M., Smith, R. F., van den Bosch, R. and Hagen, K. S. 1959. The integrated control concept. Hilgardia 29, 81-101.

Strickland, A. W. 1960. Ecological problems in crop pest control. *In* Wood, R. K. S. (ed.) Biological Problems Arising from the Control of Pests and Diseases. Symp. Inst. Biol. 9, 1-15.

Takahashi, F. 1964. Reproduction curve with two equilibrium points: a consideration on the fluctuations of insect population. Res. Pop. Ecol. 6, 28-38.

Taylor, L. R. 1965. Flight behaviour and aphid migration. Proc. N. cent. Brch. E.S.A. 20, 9-19.

Taylor, T. H. C. 1935. The campaign against *Aspidiotus destructor* Sign. in Fiji. Bull. ent. Res. 26, 1-102.

Taylor, T. H. C. 1937. The biological control of an insect in Fiji: an account of the coconut leaf-mining beetle and its parasite complex. Imperial Institute of Entomology, London.

Thorsteinson, A. J. 1960. Host selection in phytophagous insects. Ann. Rev. Ent. 5, 193-218.

Tothill, J. D., Taylor, T. H. C. and Paine, R. W. 1930. The coconut moth in Fiji: a history of its control by means of parasites. Imperial Institute of Entomology, London.

Turnbull, A. T. and Chant, D. A. 1961. The practice and theory of biological control of insects in Canada. Can. J. Zool. 39, 697-753.

Ulyett, G. C. 1947. Mortality factors in populations of *Plutella maculipennis* Curtis (Tineidae: Lep) and their relation to the problem of control. S. Afr. Dept. Agr. Ent. Mem. 2, 77-202.

Uvarov, B. P. 1957. The aridity factor in the ecology of locusts and grasshoppers of the Old World. Arid Zone Res. 8, 164-198.

van der Vecht, J. 1933. De groote peperwants of semoenjoeng *(Dasynus piperis* China). Proefschr Rijksuniv Seiden Leyden.

Varley, G. C. 1953. Ecological aspects of population regulation. Trans. IX int. Congr. Ent. (Amsterdam 1951) 2, 210-214.

Varley, G. C. 1963. The interpretation of change and stability in insect populations. Proc. R. ent. Soc. Lond. (C) 27, 52-57.

Vite, J. P. and Gara, R. I. 1962. Volatile attractants from ponderosa pine attacked by bark beetles (Coleoptera: Scolytidae). Contr. Boyce Thompson Inst. 21, 251-274.

Voute, A. D. 1946. Regulation of the density of the insect populations in virgin forests and cultivated woods. Arch. néerl. Zool. 7, 435-470.

Watt, K. E. F. 1964. Comments on fluctuations of animal populations and measures of community stability. Can Ent. 96, 1434-1442.

Watt, K. E. F. 1968. Ecology and Resource Management. Mc-Graw Hill, New York.

Way, M. J. 1953. Studies on *Theraptus* sp. (Coreidae), the cause of the gumming disease of coconuts in East Africa. Bull. ent. Res. 44, 657-667.

Way, M. J. 1954. Studies on the association of the ant *Oecophylla longinoda* (Latr.) with the scale insect *Saissetia yanyibarensii* Williams. Bull. ent. Res. 45, 113-134.

Way, M. J. 1963. Mutualism between ants and honeydew-producing Homoptera. Ann. Rev. Ent. 8, 307-344.

Way, M. J. 1966. The natural environment and integrated methods of pest control. J. appl. Ecol. 3, (suppl.) 29-32.

Way, M. J. 1968. Intra-specific mechanisms with special reference to aphid populations. *In* Southwood, T. R. E. (ed.) Insect Abundance, R. ent. Soc. Lond. Symp. 4, 18-36.

Way, M. J. and Banks, C. J. 1967. Intra-specific mechanisms in relation to the natural regulation of numbers of *Aphis fabae* Scop. Ann. appl. Biol. 59, 189-205.

Way, M. J. and Banks, C. J. 1968. Population studies on the active stages of the black bean aphid *Aphis fabae* Scop., on its winter host *Euonymus europaeus* L. Ann. appl. Biol. 62, 177-197.

Way, M. J., Murdie, G. and Galley, D. J. 1969. Experiments on integration of chemical and biological control of aphids on brussels sprouts. Ann. appl. Biol. 63, 459-475.

White, M. G. 1967. Ecological and chemical control of damage by the *Phytolyma* gall bug (Hemiptera: Psyllidae) on the iroks tree (*Chlorophora* species) in Africa. Proc. R. ent. Soc. Lond. (C) 31, 43-44.

Wiegert, R. G. 1964. Population energetics of meadow spittlebugs (*Philaenus spurmarius* L.) as affected by migration and habitat. Ecol. Monogr. 34, 217-241.

Wille, J. E. 1958. El control biologico de los insectos agricolas en el Peru. Proc. X int. Congr. Ent. (Montreal, 1956), 4, 519-523.

Wood, D. L. 1962. The attraction created by males of a bark beetle *Ips confusus* (Leconte) attacking ponderosa pine. Pan. Pacif. Ent. 38, 141-145.

Zwolfer, H. 1963. Untersuchungen uber die Struktur von Parasitenkomplexen bei einigen Lepidopteren. Z. angew. Ent. 51, 346-357.

MANAGEMENT OF INSECT PESTS[1]

By P. W. Geier

Many insects compete with us for food and organic resources, while others spread disease or cause discomfort. Although the conflict between man and insects becomes more acute as man's requirements grow (25), we are no longer the helpless victims of pests, but have become increasingly violent aggressors against insect life (22). In the recent past, means have been developed to reduce drastically the injuriousness of noxious species (28, 85), largely by the process of trial and error, and most current methods of pest control[2] are still, for a great part, empirical (83).

In a variety of ways, insects can be killed, maimed, or otherwise rendered harmless in large numbers at relatively low cost (69). Also, we know that insect numbers and their fluctuations in space and in time are the outcome of interactions between populations and their environment (64), the nature of which can be conceived and investigated analytically (24). So far, however, the two lessons have not generally been read together. We are not using to the best advantage our capabilities for harming insects, i.e., in strict accordance with what we know of the rules that govern the waxing and waning of insect numbers in nature (9). The contributions of many workers suggest that there are ways whereby pest control can be developed beyond the present empirical phase, and established in practice as an applied form of population ecology. This review attempts to present such findings in a convenient order from which an outline might be drawn of future progress in pest control.

THE NATURE OF PEST CONTROL

The origins of modern pest control are not easy to define. Present-day practices have evolved from many old and new sources, some being pragmatic and others conceptual. Attempts to present the history of pest control as a sequence of technological developments (8) are not entirely satisfying because they usually ignore the changing motivations and the conceptual progress which underlie technological advances. A comprehensive evaluation of past achievements and of current situations needs to be made in the broadest biological context.

In essence, pest control is the product of technology, defined as the study of devices to impose man's will on nature, and of biology. Conceived as a mere exercise in technology, pest control amounts to hardly more than bull-

[1] The survey of the literature pertaining to this review was concluded in December 1964.

[2] The word, "control," is used throughout in the sense of causing pest numbers to be reduced to tolerable levels.

ANNUAL REVIEW OF ENTOMOLOGY, 1966, vol. 11, pp. 471-490.

dozing nature without thought to consequences, and frequently creates more problems than it solves (11).

The contribution of biology to pest control is twofold. At the practical level, it provides the knowledge of biological facts required to devise efficient repressive techniques. At the conceptual level, as biological philosophy, it defines the general principles that underlie the evolution of pest control and may serve to guide future developments. An attempt to evaluate the use made of biological information in devising techniques of pest control is presented in Table I. In selected volumes of the *Journal of Economic Entomology*,

TABLE I

Survey of *J. Econ. Entomol.*, volumes 26, 27, 36, 37, 46, 47, 56, 57, comparing frequencies of papers dealing with problems of insect pest control (a) from an essentially technological angle, and (b) from a broader biological viewpoint.

Year of Publication	Practical control of pests				Other matters	
	Essentially technological		Substantially biological			
		% of total		% of total		% of total
1933	69	27.8	71	28.7	108	43.5
1934	69	29.5	74	31.6	91	38.9
1943	96	37.2	74	28.7	88	34.1
1944	166	53.7	69	22.3	74	24.0
1953	162	46.0	92	26.1	98	27.9
1954	143	41.8	104	30.4	95	27.8
1963	92	28.7	122	38.0	107	33.3
1964	107	25.2	169	39.8	149	35.0

the yearly percentages of papers on pest control in which a purely technological approach was adopted are compared with those of papers on similar subjects in which biological facts had been given substantial and explicit consideration. Although the amount of biology put into pest control appears to have been relatively high at all times, the survey suggests that a trend in emphasis placed on biological considerations has gone full circle from the early 'thirties to the early 'sixties. As the 'sixties, the 'thirties seem to have been a time of dissatisfaction with the current, mainly chemical, methods of control, when particular attention was given to biological evidence in the search for more satisfactory methods. The advent of organosynthetic insecticides in the late 'thirties contributed greatly toward committing insect pest control to overwhelmingly technological solutions for the next twenty years (5).

The cyclic trend apparent in Table I reflects the dual origin of pest control, i.e., its development as a "two-strand" science (32). As capabilities and knowledge temporarily grow faster in one field, practical applications draw more heavily on them, at the expense of possible contributions from the other discipline. The fact that recent shifts in emphasis could have been so strong indicates that they occurred in the absence of a unifying doctrine whereby problems in pest control could be immediately related to general biological principles. This is not to say that guiding concepts for pest control have not been proposed. Valuable contributions to the theoretical background of pest control were made quite early, e.g., Woodworth's (86) remarkable analysis and discussion of biological control, and repeatedly since, but they have not yet been fused into a single, coherent body of thought which applied workers could recognize as a common basis of reference.

THE CONCEPT OF PEST MANAGEMENT

During the past fifteen years, criticism of pest control procedures has become more searching than ever before. Not only was concern expressed about certain types of pesticides (63), or their modes of application (57), but, to an increasing degree, about the prevalent approaches to pest control which were denounced as short-sighted (56), wasteful (65), and generally ill-conceived (23). The discussion finally erupted into open controversy in the United States. Since 1962, American literature on the subject has presented a succession of charges (22) and countercharges (82), some of which are valid and dispassionate, others coloured by emotion.

It is doubtful whether pest control could have evolved along other, more satisfactory lines than those which it actually followed. The knowledge available prior to the last World War was clearly too insufficient at all levels to have made substantially better methods possible. During the war, and right up to the early 1950's when food production and public health had been restored to prewar standards (58), there was understandably little time and few resources available for large-scale experimentation. Under prevailing conditions, there was no option but to apply the bulldozing techniques which had now become enormously effective following the breakthrough in the manufacture of insecticides (5).

In recent years, as research scientists and applied workers found more opportunities for intensive investigations, the shortcomings of the bulldozer approach were gradually defined and analyzed. New ways were advocated (48) and pioneered (52). But the complexity of the subject is such that there has not been time enough to put those findings into general practice. However, an immediate purpose was served: the American controversy finally demanded an authoritative assessment of the situation, which was produced from published evidence by the Science Advisory Committee (4) appointed by the President of the United States.

The Committee's conclusions indicate clearly that the issue is not with pest control as such. The Committee finds fault with certain types of pesticides and with the excessive use made of them, because they can produce resistance to toxicants in pest populations, hazards to human health and welfare, and undesirable side effects on nontarget animals and plants. The Committee enumerates some elementary qualities required to make control procedures safer and more specific, and submits recommendations on administrative policies, *ad hoc* research, and regulatory legislation. It lists a number of ways in which noxious species have been controlled without chemical pesticides, but provides no general principles to guide the development and wider application of those methods.

Hazards to human health result from the nature of the toxicants used for control, and could therefore be eliminated, or at least strongly reduced, by appropriate improvements in the technology of insecticides and of their means of application. The inducement of resistance and the undesirable side effects on nontarget organisms, however, seem to be more deeply rooted in the chemical approach to pest control. In certain pests (44), resistance has been shown to develop against a variety of compounds in rapid succession, and the toxic action of the best narrow-spectrum insecticides such as ryania is not restricted to single species (39). It is doubtful, therefore, whether mere changes in the chemistry of insecticidal agents, or in the techniques of blanket applications could be sufficient to produce the radical improvements required. Thus, progress in technology is unlikely to provide the complete answer in many instances. It is but one element in what must amount ultimately to a new strategy of insect pest control (9).

The greatest single quality which the new strategy should possess is "selectiveness" in its action on specific populations. This selective action must remain at all times commensurate with the populations' potential injuriousness. To achieve that, the new approach must aim above all at making control operations fit well with the biology of the noxious species. Geier & Clark (37) have called this conception of pest control "protective management of noxious species," or "pest management" for short.

The idea of pest management is not original. It involves principles that have been defined many times before, e.g., by Townsend (78). Yet, there has recently been such a change of outlook as to create an intellectual climate which can be said to open a new era in pest control. The term "pest management" is meant to acknowledge that changed outlook. The term has no other value than that of a convenient label. It was coined to emphasize the comprehensive nature of the approach, and to underline its preoccupation with ecological realities. It is intended to convey the idea of intelligent manipulation of nature for man's lasting benefit, as in "wildlife management" (1).

Pest management implies acceptance of the continued existence of potentially harmful species, albeit at tolerable levels of abundance. This is not to say that pests should not be exterminated whenever feasible (50), because well-organized attempts at eradication are apt to provide the ultimate solution to the problem of noxious species. In the present state of our knowledge and capabilities, however, control by extermination can reasonably be attempted only under quite exceptional conditions. Because of this, and because successful eradication must ultimately comply with the same imperatives as any other kind of dominance over noxious species, it is taken here as a particular case in pest management. Similarly, regulatory and legislative measures of pest control are regarded solely as aids in the implementation of pest management.

THE WAYS AND MEANS OF PEST MANAGEMENT

It has been customary to classify, in terms of their material nature, the means which can be brought to bear against insects. Thus, "cultural control" meant that control was sought by adapting cultural practices, such as dates and rates of sowing, to reduce the injuriousness of a pest. "Mechanical control" entailed the use of physical means to limit harmful populations. "Chemical control" meant killing insects by mass applications of toxic substances. "Biological control" referred to the use of some species of living organisms to restrict the numbers of a pest. In recent years, as the need became apparent for more comprehensive and more flexible methods to deal with complex pest situations, the term, "integrated control," gained acceptance to designate attempts at blending together the actions of chemical insecticides and those of biological control agents (11). The term, "integrated control," differs in essence from the previous nomenclature in that it refers, not to the nature and technology of means, but to a way of using them. This shift in verbal emphasis, from formal to functional qualities in defining procedures of pest control, is in itself significant of current trends. However, the conceptual implications of the term have not been clarified satisfactorily, and it is not always evident whether it is meant to designate an empirical superimposition or juxtaposition of techniques (75), or a systematic attempt at creating new or modified ecological systems in which pest populations can be stabilized at tolerable levels of abundance. Most commonly, "integrated control" is intended to imply some degree of discrimination in the conventional use of pesticides (68).

In pest management, the means available for control are considered primarily according to their functional qualities which are twofold in nature. These qualities relate first to the technical characteristics and mode of operation of the means, and to their attributes, i.e., safety, inherent selectiveness, and intrinsic biological efficacy. Second, they concern the ultimate effects of the means, i.e., the rôles which they perform in ecological systems. The present section reviews the nature and intrinsic qualities of the possible means of pest management.

Pest management seeks to deal with "populations" as defined by Huffaker & Messenger [(46), p. 77], for they are the biological units in which species exist. A population results from complex and continuing interactions between the innate qualities of a species and the effective attributes of the environment in which the population is both supported and limited numerically. The abundance of a population is a function of the fitness of the innate qualities of individuals in the population for the operative features of the environment, for example, climate and weather, and the supply and distribution of vital requisites such as food and shelter (2). This fitness can be reduced (*a*) by modifying the innate qualities which enable individuals to perform their life functions within the population and the environment; or (*b*) by altering essential features of the environment in such a way as to make it no longer able to support large numbers of a population while leaving the innate qualities of its members unchanged. It is convenient to consider the means of pest management according to which of the two kinds of effect they are meant to implement.

The aim of alternative (*a*) is to burden the population with "useless" individuals which will reduce the chances of the "useful" individuals to survive and reproduce in kind. Methods having this effect have been termed "autocidal" because the reductions in numbers achieved by them result from self-destroying changes within the individuals of a species.

The aim of alternative (*b*) is to render the environment of a population less favourable in order to lower its carrying capacity for the target species. This can be done in three ways: by increasing the frequency and severity of environmental hazards and thereby increasing the probability of premature death for individuals; by preventing individuals from locating vital requisites (such as food, shelter, mates, etc.) for the performance of their life functions; and by curtailing the supply of vital requisites to such an extent that former habitats become unsuitable for the pest species.

Autocidal Methods

So far, autocidal effects have been achieved by burdening noxious populations with sterilized individuals capable of competing for normal mates with fertile members of their own sex (53). Induced sterility can assume different forms, e.g., total inability, or reduced ability to mate; production of nonfunctional gametes entailing no loss of mating ability; or genetic alterations resulting in nonviable zygotes after fertilization (73). Impairment of the ability to mate is not desirable. Although individuals thus affected may contribute to depriving normal individuals of requisites in short supply, their numbers in a population would have to be inordinately large to depress significantly the rate of increase in an otherwise favourable environment.

Three methods are known whereby insects can be sterilized without unduly reducing their vitality or modifying their behaviour: by ionizing radiation (21); by radiomimetic and antimetabolic chemosterilants (54); and by exposure to electronic flashes (70). Ionizing radiation has been used successfully for large-scale operations (12, 43). It is a safe and selective method, but is relatively costly because it involves producing, treating, and releasing large numbers of individuals. Its application is limited economically to species with suitable biological attributes, namely, appropriate mating habits and patterns of seasonal development, adequate powers of dispersal, and harmlessness in the adult stage. Chemosterilization shows promise of greater flexibility. It could be implemented at little expense by direct action on members of the target population in nature, providing that sterilizing agents could be used in a manner that would limit their effect exclusively to the pest. Sterilization by electronic flash might conceivably serve the same purpose in some instances.

The concept of autocidal systems in pest control was further elaborated by Downes (30) who discussed the possibility of exploiting genetic incompatibilities between different races of the same species, e.g., the gypsy moth, *Porthetria dispar* (L.); or, following Vanderplank's (80) lead, between reproductively, but not sexually, isolated forms such as the tsetse flies, *Glossina morsitans* Westwood and *G. swynnertoni* Austen. Viable, but sterile, offspring could be produced in this way, which would add to the efficacy of the method by competing for requisites with normal individuals, or by increasing the adverse effects which other density-related processes may exert on the population. More recently, this approach was also considered by Craig (27) for the control of disease vectors.

Crosses with forms of the same species genetically adapted to different environmental conditions can lessen the fitness of local populations, as shown by Hogan (42) in the common field cricket, *Teleogryllus commodus* (Walker). By mating them with males of the northern Australian, nondiapausing form, females of the southern, diapausing form are forced to produce nondiapausing eggs which cannot survive the southern winter. Applications of this finding in a new approach to the control of the pest are being investigated as a promising instance of autocidal action.

The possibility of restoring susceptibility to insecticides in newly resistant populations was envisaged by Feiding (33). Such attempts would amount to a form of control in which genetic phenomena were made to play an autocidal rôle within a population, the mortality in which was due predominantly to insecticides.

In large-scale operations, autocidal action has proved to be one of the most powerful means of reducing populations of noxious insects (53). Its efficacy is due to the direct effects which autocidal interventions can exert upon large numbers of individuals; to the consequences of such effects for untreated members of the interbreeding group; to the rapidity with which those consequences can spread throughout a population; and, finally, to the fact that there are no immediate biological defenses with which individuals of susceptible species can oppose autocidal methods.

Modifications of Environmental Qualities

Hazards.—Hazards are created by mortality agents to the action of which insects are exposed in their environments. Whereas autocidal procedures can achieve an immediate effect on the birth rate of populations by using live but useless individuals, hazards cause the direct destruction of insects and thus tend to bear immediately upon the mortality rate. Hazards can result from natural components of the environment, or from artefacts which may be temporary, recurrent, ephemeral, or long-lasting. Pest management may seek to exploit existing hazards, to increase their efficacy, or to create new hazards.

Conventional insecticides and their functional homologs, e.g., certain pathogens, constitute man-made, temporary or recurrent hazards in the environment of a population. The most powerful among them are the chemical insecticides in their various forms and applications.

Extensive efforts are being devoted to making conventional pesticides safer for man and more specific in action by improving their technological qualities (17). Suggestions for new chemical insecticides include the use of antimetabolites (67); of abrasives to kill insects by lowering the resistance of their cuticle to water loss (31); of compounds capable of eliminating or interrupting prematurely the protective diapause of certain species, e.g., urea in the case of the field cricket, *T. commodus* (3, 41).

Examples of developments in the related field of physical insecticides and of pest destruction by mechanical means are: the irradiation of commodities and stored products (18); the use of ultrasound (72); certain mechanical trapping devices (8); the destruction of grain pests by percussion (7); exposure to lethal temperatures and humidities (26); exposure to artificial illumination for short critical periods to prevent the onset of protective diapause (10).

Hazards result also, of course, from the existence of inimical organisms, i.e., predators, parasites, pathogens, and interspecific competitors, in the environment of a population. The tendency has been to regard the study of such hazards and their use for pest control as a separate discipline within the general field of applied entomology [(29), p. 9]. Most inimical organisms constitute hazards whose influence on target populations does not differ essentially from that of insecticides in lessening the favourableness of environments. Their immediate efficacy against pests is frequently inferior to that of insecticides [(79), p. 699]. However, their use generally requires less intervention by man, because many inimical organisms are able to maintain themselves by reproduction, so that their effect is permanent or spontaneously recurrent. Yet, as with chemical insecticides, there are instances in which parasites like *Trichogramma* sp. (34), predators like *Rodolia cardinalis* (Mulsant) (16), and pathogens like *Bacillus thuringiensis* (Berliner) (76), are useful only in repeated applications. Biological pesticides are safer for man than chemical poisons, and usually exert fewer undesirable side effects.

Denial of access to existing requisites.—Insects find their requisites and evade certain hazards by reacting to external signals. It is possible to interfere with perception-response processes by producing misleading signals which cause the insects to fail to locate their requisites in an otherwise favourable environment. This can be done in three ways: by making attractive but unfavourable substitutes more directly accessible than normal requisites; by conferring inhibiting or repellent qualities upon normal requisites; and, finally, by confusing the perception of requisites by pests.

The first method involves the use of lures, such as baits and pheromones, in the study and application of which substantial advances are being made (14, 49, 88). Insects can be deterred by repellent or inhibiting stimuli from utilizing their normal requisites. Repellent effects have been achieved in a number of ways, e.g., by simulating acoustically the presence of bats whose echo-locating signals are perceived by approaching adults of the pest species (13). Inhibiting agents, such as "antifeeding" compounds, have proved to be effective in a number of instances and show promise of further application (87). Finally, tests are being made to determine the possibility of denying insects access to their mates, thereby preventing reproduction by subjecting them to a confusing mass of conflicting sex signals (19).

The particular value of these procedures lies in their economy, safety, and selectiveness; in the power of their compelling influence; and in the nature of the biological urges which they exploit.

Suppression of requisites.—It is possible to modify intrinsically favourable habitats in such a way that they no longer provide adequate environments for certain populations. This is done by withholding or withdrawing, either temporarily or permanently, some essential requisite of the target species. For instance, appropriate cultural techniques can break the continuity of a pest's food supply (62), or destroy indispensable shelter (59). Many other methods follow the same principle, for example, Mules' operation to protect sheep from blowfly strike (61) or in the use of airtight grain containers to kill contaminating insects by anoxia (6).

The ultimate development of this approach consists of selecting or creating forms of a normally susceptible plant or animal, which are intrinsically unsuitable as sources of food for a pest. This has been achieved with a degree of efficacy which ranges from useful in particular circumstances, as with the apple variety, Northern Spy, with respect to the woolly aphis *Eriosoma lanigerum* (Hausmann) (71), to being practically complete, as with the American vines used as rootstocks to protect the European grapevine against the grape phylloxera, *Viteus vitifoliae* Fitch (47). The approach obviously holds great promise (66) but its implementation is particularly slow—mainly because of the difficulties involved, not the least of which is a general lack of interdisciplinary communication between entomologists and plant and animal breeders.

The means of pest management differ greatly in their functional qualities. Some are self-perpetuating, as are many biological agents for instance; others, such as insecticidal sprays and dusts, need to be applied at frequent intervals for lasting protection. Still others occupy intermediate positions between these extremes. Certain processes, such as those involving the use of lures, may have to be kept in operation continually, but lend themselves to infrequent interventions with long-lasting effects.

A difference of another kind is found in the ways in which the means of pest management affect the rate of increase of target populations. With autocidal methods, the event of sterilizing one individual can immediately prevent reproduction by others, and tends to depress the birth rate directly. The interventions involved in autocidal methods may thus be said to produce multiple-stage effects which can automatically depress the rate of increase at all levels of population density. The demographic effects of other means of control are not inherently so far-reaching. In general, means designed to lessen the favourableness of the environment exert, in themselves, no direct multiple-stage action, and the consequences for the population of the immediate mortality which they induce are largely dependent upon the conditioning influences resulting from other circumstances such as population density and the operations of subsidiary repressive agents.

Most control methods depend for development and application upon the existence of a number of diversely oriented scientific institutions. Some require elaborate supporting industries, whereas others need little or no technological capability. Finally, control methods vary greatly in feasibility and in optimum scale of application. Some, involving, for instance, the use of conventional pesticides and cultural practices, lend themselves to limited action in time and space for quick protection and for application by individual users. Others, like autocidal methods, are best implemented in extensive campaigns organized and carried out by specialists. In their diverse qualities, the means of pest management cannot be rated according to inherent usefulness or absolute value. None of them can ever provide a cure for all pests, but all possess some particular merit. One must conclude with a truism: for satisfactory performance, they need to be used in the circumstances for which each is best suited. Those circumstances, however, are not obvious, nor can they be defined precisely in the present state of our knowledge. An attempt is made in the following section to outline some principles which may help to conceive them.

THE PRACTICE OF PEST MANAGEMENT

The success of pest control operations is judged both by results and by cost. The best method of control is that which gives adequate relief from a pest for the lowest initial outlay and recurrent expenditure of effort and money. Pest management seeks to realize such an optimum. However, in the present state of our knowledge, there is no way to define an absolute optimum for each and every pest situation. The value of control procedures changes in time and in space with such variables as the progress of biological understanding, the development of technological capabilities, changing natural circumstances, and differences in the thresholds of pest tolerance.

DeBach [(29), p. 675] observed that success or failure in biological control could not be related *a priori* to single attributes of target species, of destructive agents, or of environments—or to any particular technique of application. His observation applies to all forms of pest control. Highly successful procedures have been devised in a diversity of circumstances, using a diversity of means to reduce a great variety of noxious life forms. It appears therefore that pest control can be achieved by methods involving any of the means discussed in the foregoing section if these are used to the best advantage.

In order to ensure that control methods are used to best advantage, it is necessary to have an understanding of the natural influences which determine the abundance and persistence of insect populations. In other words, a knowledge of the population dynamics of insects is relevant to the practice of pest management. The following simple model, which is derived from Tansley's (77) flexible concept of "ecosystem," provides an adequate orientation for understanding the population dynamics of a particular pest species (Fig. 1).

For convenience of study and argument, Clark (24) and Geier (36) have proposed that the ecosystem be regarded as a complex of overlapping subsystems, each erected around one of the single-species populations incor-

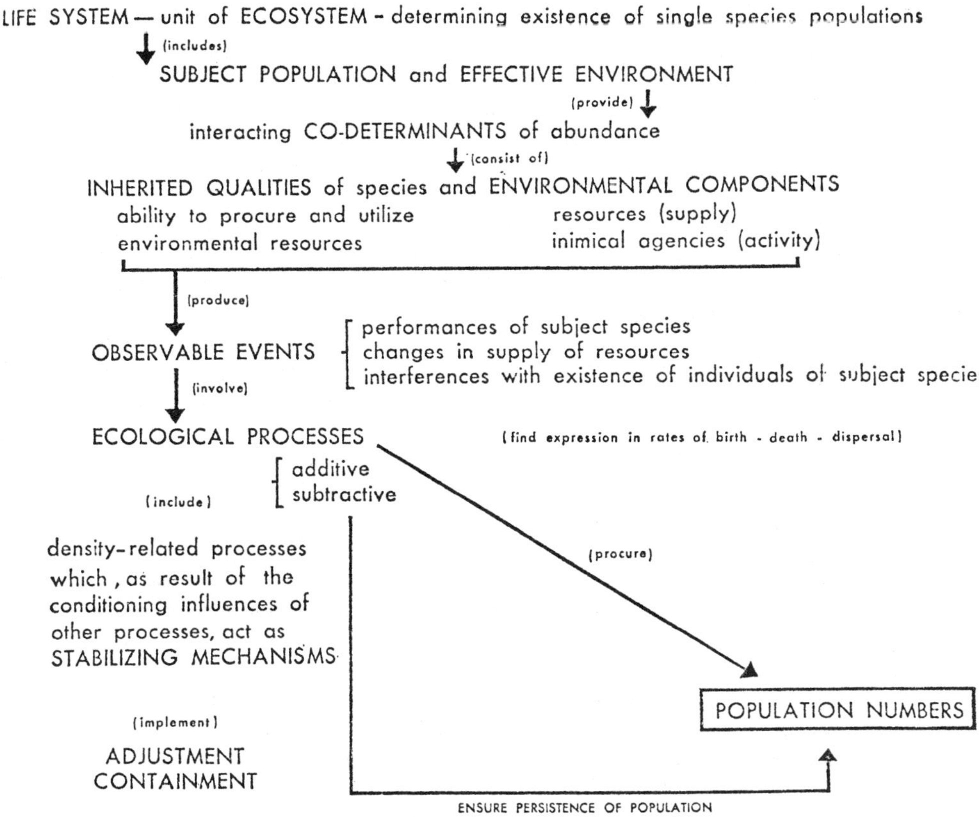

FIG. 1. Diagrammatic summary of the functional concept of life system.

porated functionally in the ecosystem. Those subsystems were called "life systems," and the species to which each refers, the "subject species." Thus, life-systems comprise only those components of an ecosystem which affect the population(s) of the subject species significantly. Populations result from the continuing interactions between the individual and collective qualities of a species and the operative attributes of the environment. The numbers of a species in an area are the result of many kinds of interactions between those basic "co-determinants" of abundance, interactions which are observable in the form of ecological events and processes. The persistence of populations involves the operation of negative feedback mechanisms (74) which bring about the "stabilization" of population numbers, in the sense of preventing increase to extinction, that is, to levels that would necessarily entail catastrophic mortality. Such stabilizing mechanisms may, for example, be provided by intraspecific competition, by emigration due to overcrowding, or by the density-dependent destructive action of inimical agencies such as certain parasites and predators. Intraspecific competition and other density-related forms of mutual interference between individuals of a subject population serve to stabilize population numbers by "adjusting" density to the availability of a limiting requisite, whereas suitable destructive agents serve to stabilize population numbers by "containing" them at densities below those which the supply of requisites such as food and shelter would otherwise support. Whether or not ecological processes resulting in density-related interference between individuals, or in density-dependent destruction, actually function as the stabilizing mechanism of a subject population depends upon the "permissive" or "conditioning" effects of other events and processes in the life system, which also contribute toward determining the population's abundance (46).

Adjusting and containing mechanisms are not equivalent in their stabilizing effect. The first operate without fail, repeatedly and uniformly, for they are activated solely by the abundance of the subject population relative to the availability of a requisite in short supply. They function in a wholly "deterministic" way. The processes which operate as containing mechanisms are essentially "probablistic" in effect, because their function depends not only upon the densities of the subject population, but also upon many other variables which can influence the action of the agent concerned independently of the subject population. However, the chance element can be minimized in the actual operation of those processes, e.g., by increasing the number of agents in the life system that contribute effectively in implementing them (20).

Stabilizing mechanisms are of the utmost importance in pest management, because it is exclusively through their existence that reductions in population numbers achieved by insecticidal means of all kinds can be consolidated into lasting repression. The means of pest control are not density-stabilizing per se. In order to allow stabilization to occur, they must fit into the life system of target populations in such a way as to create conditions which permit new stabilizing mechanisms to take over, or at least to permit or condition existing mechanisms to operate effectively at lower densities of the subject population than was previously the case.

From the foregoing considerations, the practice of pest management can be defined as consisting of three phases: (*a*) Determining how the life system(s) of a pest need(s) to be modified for the numbers of the species to be reduced and held at tolerable levels. This can be done empirically, or by systematic study. (*b*) Finding how the desirable modifications can be obtained in view of the biological knowledge and the technological means available. (*c*) Devising the procedures of control best suited to current possibilities and requirements, in which available means are made, if possible, to serve as conditioning or stabilizing agents in modified life systems.

The practice of pest management is best illustrated by reference to concrete examples. In Table II, an attempt is made to classify the basic procedures of pest control according to (*a*) their intrinsic value, and (*b*) the relative frequency of the interventions needed to implement them against persistent populations. The intrinsic value of a control procedure is defined as the degree of assurance offered that its use will regularly and lastingly achieve the measure of protection required.

TABLE II

ECOLOGICAL SYNOPSIS OF PROCEDURES FOLLOWED TO REDUCE THE INJURIOUSNESS[a] OF NOXIOUS INSECTS

Expected need for renewed interventions		Intrinsic value of control procedures: A. Unsatisfactory	B. Acceptable	C. Good
	Nil	7. Use of permanent, but unreliable, mortality agents resulting in undependable containment	8. Use of permanent mortality agents, reliable singly or in combination, to produce dependable containment	9. Measures to curtail the supply of requisites of pests
	Periodic	4. Intermittent interventions to supplement undependable density-containing mechanisms	5. Intermittent interventions to restore conditions for effective containment	6. Use of devices to divert pests from, or deny them access to, requisites in permanent supply
	Continuing	1. Arbitrary use of nonpersistent mortality agents to reduce injurious populations	2. Systematic use of nonpersistent mortality agents to contain potentially injurious populations	3. Use of nonpersistent means to suppress or mask requisites in permanent supply

[a] The original favourableness of environments is taken to remain unaltered but for the effects of the control procedures. Measures destined primarily to increase the tolerance of host organisms to pest attack are not considered specifically.

Procedure 1 is essentially curative. It seeks to reduce populations after they have reached dangerous numbers. The timing of insecticidal operations of this sort makes no allowance for the possibility that all stages of the pest may not be equally susceptible to treatment, and that different developmental stages may coexist. Also, sprays applied only in parts of the habitat of a wide-ranging pest have little chance of preventing the rapid reinfestation of treated areas. Curative measures arbitrarily applied are justified in emergencies, but they cannot constitute a satisfactory control procedure in the long run because of their uncertain efficacy, the likelihood of inducing resistance to pesticides, the possible danger to humans, and undesirable side effects on other species. Procedure 1 is the least desirable, as may be seen from the results obtained by its use against the indigenous leafroller, *Epiphyas postvittana* (Walker), in the apple orchards of southeastern Australia. As a consequence of indiscriminate applications of broad-spectrum insecticides during the past 15 years, the species has acquired a considerable degree of resistance to chlorinated hydrocarbons and, in certain instances, to organo-phosphorus compounds. Seasonal control programmes, including as many as 12 sprays of various pesticides, have recently failed to give commercial control of the pest in some areas, and no conventional toxicant shows any promise of improving the efficacy of the present practice (unpublished data).

In contrast, Procedure 2 is essentially preventive. It consists of adapting the use of conventional insecticides to the biology of the target species so as to cause maximum mortality and thus prevent pest populations from reaching dangerous numbers. This approach is taken with the so-called "planned dipping" of cattle in acaracides to destroy the tick *Boophilus microplus* (Canestrini) in northern Australia (38). To be effective, dipping must occur at sufficiently short intervals to prevent the surviving ticks from reproducing and thereby causing the reinfestation of treated cattle. Although simple and relatively efficient, the procedure has drawbacks. In addition to requiring strict adherence to timing, it presents some risk of inducing resistance to pesticides in the tick and can entail undesirable side effects. However, those disadvantages are not inherent in the procedure, but in the means used to implement it.

In Procedure 3, a requisite of the species is withdrawn repeatedly in otherwise favourable environments. Such, for instance, is the effect of "pasture spelling," a methodical rotation of grazing areas recommended in northern Australia to avoid the build-up of injurious populations of cattle tick (84). This procedure is simple and highly effective. Besides being safe and entailing no major side effects, it leaves the pest species no possibility of adaptive defense. However, suppressive procedures of this kind tend to complicate routine husbandry and are therefore not always readily adopted.

Procedure 4 is illustrated by the attempts of Hinds & Spencer (40) to control the sugarcane borer, *Diatraea saccharalis* (Fabr.), in Louisiana by periodically supplementing natural populations of the egg parasite, *Trichogramma minutum* Riley, with artificially reared individuals. The parasite had been recognized as an important limiting agent of the pest, but it failed to become sufficiently abundant to exert its full effect before the latter part of the growing season. Accordingly, large numbers of *T. minutum* were released in the spring to promote early multiplication in areas infested by the pest. Although some measure of success is claimed to have been achieved, it was evident that the conditions which frequently prevailed at the beginning of the season were such as to prevent the egg parasites from operating effectively. To ensure adequate protection against the borer under all conditions, *T. minutum* would probably have to be bred and released in such numbers as to make the procedure impractical.

In Procedure 5, insecticidal means are used in such a way as to allow other agencies to prevent or retard a subsequent build-up of treated populations to injurious numbers. This appears to have been achieved in the method adopted to protect forest stands from a fatal degree of defoliation during the recent outbreak of the spruce budworm, *Choristoneura fumiferana* (Clemens), in New Brunswick (60). Stands in which the pest was found to have reached densities that threatened the survival of balsam fir were sprayed from the air with small quantities of DDT. In seven years' operations, nearly half of the sprayed areas did not require a second treatment (81); and, beside inducing no other species to increase to injurious numbers the insecticidal treatments did not interfere with the processes which ultimately caused the collapse of the spruce budworm outbreak (55). It must be inferred that these effects resulted from the operations of density-stabilizing mechanisms which almost certainly involved parasites and predators whose action on the subject population was unimpaired and, in some instances, even increased after applications of the toxicant. That processes capable of containing budworm populations at low densities do operate after numbers have been reduced with chemical insecticides is confirmed by Furniss' (35) findings in the Pacific Northwest. Although the method applied in New Brunswick proved to be highly effective, it has the inherent disadvantage of requiring a very precise understanding of critical pest situations. The risks of inducing insecticide resistance in the target species and of adversely affecting the fauna of the treated areas is not a characteristic of the procedure, but of the insecticidal means and techniques of application used to implement it.

Procedure 6 consists basically of preventing individuals of the pest species from gaining access to requisites such as food, shelter, or mates. As shown above, this can be achieved by making attractive, but unfavourable, substitutes more readily available than the normal requisite. If such substitutes are rendered lethal to the pest individuals that come in contact with them, they provide a very effective method of control capable of operating automatically over long periods. Such procedures have been applied with success in several instances, notably against the Queensland fruit fly, *Strumeta tryoni* (Froggatt), in New South Wales (M. A. Bateman, unpublished data). In this case, two chemical lures, a derivative of phenylbutanone which is attractive to males only, and a solution of hydrolyzed protein attractive to both sexes, were used in combination with insecticides to protect whole localities. The phenylbutanone derivative was absorbed, together with a toxicant, onto small blocks of fibreboard which were nailed in protected places on trees and fences. The protein hydrolysate was spot sprayed with an insecticide on foliage at intervals along roads. The protein hydrolysate proved to be more effective than the male attractant, but combined applications of the two poisoned lures gave complete control of the pest in all circumstances. Besides being highly effective, this procedure presents a reduced risk of inducing insecticide resistance in the target species, its effects are specific, and it can be applied without danger to humans. Against wide-ranging pests, however, it needs to be implemented under the authority of a coordinating body.

In Procedure 7, a permanent hazard is successfully created in the environment of a pest population by means of an inimical organism, thereby causing the numbers of the pest to be contained at lower mean densities. However, the qualities of the inimical organism may be insufficient to ensure in all circumstances the numerical stabilization of the target population at low levels of abundance, and, from time to time, outbreaks occur. The extent to which Procedure 7 is likely to fail temporarily is determined by the degree in which the stabilizing effect induced by the mortality agent is conditioned by events unrelated to the densities of the subject population (46). This eventuality is illustrated in Huffaker, Kennett & Finney's (45) study of *Aphytis maculicornis* (Masi) as an imported parasite for the control of the olive scale, *Parlatoria oleae* (Colvée), in California. While normally able to multiply more rapidly than its prey in winter and thus reducing it to negligible numbers by spring, the parasite is ill-adapted to extreme summer conditions and may sometimes be so scarce in autumn as to exert no significant influence on scale numbers during the following winter.

Procedure 8 results from the establishment in a pest's environment of several inimical agencies capable of functioning collectively to contain the subject population at permanently tolerable levels of abundance. An example is provided by the European spruce sawfly, *Diprion hercyniae* (Hartig), in eastern Canada (15). The pest appeared there at the turn of the century, increasing in abundance until limited by the supply of foliage. Between 1934 and 1939, large numbers of natural enemies were imported, of which seven became established, and two, *Exenterus claripennis* Thomson, a larval parasite, and *Dahlbominus fuscipennis* (Zetterstedt), a pupal parasite, were considered to exert considerable control by 1938. But a virus disease, probably introduced with the imported material, appeared in the field in 1938, and by 1942 had become distributed throughout the area infested by the sawfly, causing the numbers of the pest (which had remained relatively very high despite the action of the released parasites) to crash to insignificant levels. There followed a period of very low sawfly densities until 1944, when an upsurge occurred, linked with a disappearance of the virus. The outbreak was checked largely through the action of two imported parasites, *Drino bohemica* Mesnil, and *Exenterus vellicatus* Cushman, which had not been previously noted for their efficacy and which had survived the virtual disappearance of their host following the epizootic. Although the pathogen remained endemic and was later found to be sporadically active, those species "appear to have been chiefly responsible for maintaining the sawfly populations at an extremely low level until 1951 when, for an unknown reason, the sawfly suddenly increased in numbers. The control of this [incipient] outbreak is more obviously due chiefly to an increased activity of the parasites" (15). The fact that injurious defoliation by *D. hercyniae* could be prevented for the last 20 years is proof of the reliability of control procedures which compound the effects of several agencies whose operations are not exclusively governed by the densities of target populations. Such situations are not exceptional in nature: combined actions by predators, parasites, and pathogens are generally considered to be instrumental in the permanent scarcity of most insect species (64). It is likely that beneficial changes similar to those induced in the Canadian life system of the European spruce sawfly by the introduction of effective, coacting or complementary repressive agents could be implemented in the life systems of many other pests.

Finally, in Procedure 9, a pest problem is solved in the neatest and most definitive way be suppressing both its existence and the likelihood of its recurrence. Such an instance is found in the breeding of cattle resistant to the tick *B. microplus* for use in northern Australia [(51), p. 235]. It is clear, however, that such situations are not readily created, and must therefore be regarded as exceptional achievements in pest control.

The control procedures of Type A (first column Table II) are basically unsatisfactory because of their uncertain efficacy. The corresponding procedures of Type B (second column) are improvements on them. Procedures of Type B aim to contain the population densities of pests at levels below those which the supply of such environmental requisites as food and favourable living space would permit. In Procedures 5 and 8, considerable benefit is derived from the effective operation of stabilizing processes, which are not provided by Procedure 2. Here, a similar effect must be achieved by repeatedly destroying very large fractions of the target populations through direct action. In the present state of our capabilities, the need for adopting Procedure 2 rather than 5, and 5 rather than 8, often arises from the fact that the first-mentioned procedures are generally better suited than the second to maintain the low densities of pests required in the commercial protection of intensive crops.

All procedures of Type C (third column) tend to impose on target populations the need to adjust their numbers either to the restricted availability (Procedures 3 and 6) or to the reduced supply (Procedure 9) of some vital requisite. Their particular value lies in the constancy of their effect, itself resulting from the automatic operation of the stabilizing mechanisms which they induce. Those procedures may naturally affect the local extermination of pests, but their feasibility is limited for biological and practical reasons. This is particularly obvious in situations which require control not only of one but an array of potentially noxious species, as in the defense of highly susceptible crops. In such instances, there is usually no option but to resort to the procedures of Type B.

CONCLUSION

The procedures of Type B are the most immediately and generally applicable forms of pest management. Such procedures can be developed in all manners of ways, using whatever means are best suited to increase the resistance or tolerance of the organisms exposed to pest attack, and to reduce lastingly the abundance of pest species through the action of effective and dependable stabilizing mechanisms. The need to manage several pests at once must ultimately lead to complex programmes calling for the genuine integration—and not merely the juxtaposition—of different control procedures, which would amount to manipulating the whole ecosystem rather than single life systems. This, the ultimate aim of pest management, has yet to be attempted.

Because man must still depend largely upon natural resources for the supply of his own requisites, the future solution to pest problems is more likely to be a matter of learning to live with potentially noxious species than a matter of doing away altogether with unwanted organisms. Thus, progress in the control of pests appears to depend primarily on a greatly increased knowledge of the population ecology of noxious, or potentially noxious, species in order to define adequately their status as pests, to predict significant changes in their numbers, and to manipulate their life systems effectively; and, secondarily, on the technological refinement of known insecticidal means, or the development of new insecticidal procedures.

LITERATURE CITED

1. Alexander, M. M. The place of ageing in wildlife management. *Sci. Am.* **46,** 123–37 (1958)
2. Andrewartha, H. G., and Birch, L. C. *The Distribution and Abundance of Animals* (Univ. of Chicago Press, Chicago, 782 pp., 1954)
3. Anon. *Rept. Biol. Br. Vict. Dept. Agr., No. 2,* 42 (1963)
4. Anon. Use of pesticides. *Rept of President's Sci. Advisory Committee,* 1–25 (1963)
5. Arrington, L. G. World consumption of pesticide materials. *Agr. Chem.,* **6,** 53–57, 85 (1951)
6. Bailey, S. W. Air-tight storage of grain; its effects on insect pests. I. *Calandra granaria* L. *Australian J. Agr. Res.,* **6,** 33–51 (1955)
7. Bailey, S. W. The effects of percussion on insect pests of grain. *J. Econ. Entomol.,* **55,** 301–4 (1962)
8. Balachowsky, A. S. *La lutte contre les insectes* (Payot, Paris, 382 pp., 1951)
9. Balch, R. E. The future in forest entomology. *J. Forestry.* **62,** 11–18 (1964)
10. Barker, R. J., Cohen, C. F., and Mayer, A. Photoflashes: A potential new tool for control of insect populations. *Science,* **145,** 1195–96 (1964)
11. Bartlett, B. R. Natural predators. Can selective insecticides help to preserve biotic control? *Agri. Chem.,* **11,** 42–44, 107 (1956)
12. Baumhover, A. H., Graham, A. J., Bitter, B. A., Hopkins, D. E., New, W. D., Dudley, F. H., and Bushland, R. C. Screw-worm control through release of sterilized flies. *J. Econ. Entomol.,* **48,** 462–66 (1955)
13. Belton, P., and Kempster, R. H. A field test on the use of sound to repel the European corn borer. *Entomol. Exptl. Appl.,* **5,** 281–88 (1962)
14. Beroza, M., and Green, N. Synthetic chemicals as insect attractants. *Adv. Chem. Ser.,* **41,** 11–30 (1964)
15. Bird, F. T., and Elgee, D. E. A virus disease and introduced parasites as factors controlling the European spruce sawfly, *Diprion hercyniae* (Htg.), in central New Brunswick. *Can. Entomologist,* **89,** 371–78 (1957)
16. Bodenheimer, F. S. *Icerya purchasi* Mask. und *Novius cardinalis* Muls. *Z. Angew. Entomol.,* **19,** 514–43 (1932)
17. Brinkley, P. C. Pesticides and the pesticide industry. *Bioscience,* **14,** 26–28 (1964)
18. Bull, J. O., Wond, T., and Cornwell, P. B. A comparison of the susceptibility of the grain weevil (*Sitophilus granarius* L.) to accelerated electrons and ^{60}Co gamma radiation. *Atomic Energy Research Establishment Rept. No. 3890,* 1–45 (1961)
19. Burgess, E. D. Gypsy moth control. *Science,* **143,** 526 (1964)
20. Burnett, T. Interactions in insect populations. *Am. Naturalist,* **94,** 201–11 (1960)
21. Bushland, R. C., and Hopkins, D. E. Sterilization of screw-worm flies with X-rays and gamma rays. *J. Econ. Entomol.,* **46,** 648–65 (1953)
22. Carson, R. *Silent Spring* (Hamish Hamilton, London, 304 pp., 1962)
23. Chant, D. A. Strategy and tactics of insect control. *Can. Entomologist,* **96,** 182–201 (1964)
24. Clark, L. R. The population dynamics of *Cardiaspina albitextura. Australian J. Zool.,* **12,** 362–80 (1964)
25. Cole, L. C. Pesticides: a hazard to na-

ture's equilibrium. *Am. J. Public Health*, **54,** 24–31 (1964)

26. Cotton, R. T. *Pests of stored grain and grain products.* (Burgess Publ. Co., Minneapolis, Minn., 306 pp., 1941)
27. Craig, G. B., Jr. Prospect for vector control through genetic manipulation of populations. *Bull. World Health Organ.*, **29,** 89–97 (1963)
28. Darby, W. J. Pesticides: A contribution to agriculture and nutrition. *Am. J. Public Health*, **54,** 18–23 (1964)
29. DeBach, P., Ed. *Biological Control of Insect Pests and Weeds* (Chapman & Hall, London, 844 pp., 1964)
30. Downes, J. A. The gypsy moth and some possibilities of the control of insects of genetical means. *Can. Entomologist*, **91,** 661–64 (1959)
31. Ebeling, W. Physicochemical mechanisms for the removal of insect wax by means of finely divided powders. *Hilgardia*, **30,** 531–61 (1961)
32. Egler, F. E. Pesticides—in our ecosystem. *Am. Scie.*, **52,** 110–36 (1964)
33. Feiding, J. Possible reversal of resistance. *Bull. World Health Organ.*, **29,** 51–62 (1963)
34. Flanders, S. E. Mass production of egg parasites of the genus *Trichogramma. Hilgardia*, **4,** 465–501 (1930)
35. Furniss, R. L. Answers to some criticisms of aerial spraying in forest management. *J. Forestry*, **57,** 260–62 (1959)
36. Geier, P. W., Population dynamics of codling moth, *Cydia pomonella* (L.) in the Australian Capital Territory. *Australian J. Zool.*, **12,** 381–416 (1964)
37. Geier, P. W., and Clark, L. R. An ecological approach to pest control. *Proc. Tech. Meeting Intern. Union for Conserv. of Nature and Natural Res., 8th, 1960, Warsaw*, 10–18 (1961)
38. Harley, K. L. S., and Wilkinson, P. R. A comparison of cattle tick control by "conventional" acaricidal treatment, planned dipping, and pasture spelling. *Australian J. Agr. Res.*, **15,** 841–53 (1964)
39. Heal, R. E. Ryania insecticides. *Agr. Chem.*, **4,** 37–90 (1949)
40. Hinds, W. E., and Spencer, H. *Trichogramma* experiments in 1928 for control of the sugarcane borer. *J. Econ. Entomol.*, **22,** 633–36 (1929)
41. Hogan, T. W. The action of urea on diapause in eggs of *Acheta commodus* (Walk.) *Australian J. Biol. Sci.*, **14,** 419–26 (1961)
42. Hogan, T. W. Some diapause characteristics and interfertility of three geographic populations of *Teleogryllus commodus* (Walk.) *Australian J. Biol. Sci.* (In press)
43. Horber, E. Maikäfer gegen Maikäfer. *Mitt. Schweiz. Landwirtsch.*, **11,** 145–55 (1963)
44. Hoskins, W. M. Resistance to insecticides. *Intern. Rev. Trop. Med.*, **2,** 119–74 (1963)
45. Huffaker, C. B., Kennett, C. E., and Finney, G. L. Biological control of olive scale, *Parlatoria oleae* (Colvée), in California by imported *Aphytis maculicornis* (Masi). *Hilgardia*, **32,** 541–636 (1962)
46. Huffaker, C. B., and Messenger, P. S. The concept and significance of natural control. In *Biological Control of Insect Pests and Weeds*, Chap. 4, 74–117 (DeBach, P., Ed., Chapman & Hall, London, 844 pp., 1964)
47. Husmann, G. C. Testing Phylloxera-resistant stocks in the vinifera regions in the United States. *U. S. Dept. Agr. Tech. Bull.*, **146,** 1–54 (1930)
48. Jacob, F. H. Some modern problems in pest control. *Sci. Progr.*, **46,** 30–45 (1958)
49. Jacobson, M. Recent progress in the chemistry of insect sex attractants. *Advan. Chem. Ser.*, **41,** 1–10 (1964)
50. Jenkins, C. F. H. Insect pests and their control. *J. Agr. W. Australia*, **2,** 225–38 (1961)
51. Kelly, R. B. *Native and Adapted Cattle* (Angus & Robertson, Sydney, Australia, 275 pp., 1959)
52. Knipling, E. F. Sterile-male method of population control. *Science*, **130,** 902–4 (1959)
53. Knipling, E. F. The potential role of the sterility method for insect control with special reference to combining this method with conventional methods. *U. S. Dept. Agr., ARS*-33–98 (1964)
54. LaBrecque, G. C. Chemosterilants for the control of houseflies. *Advan. Chem. Ser.*, **41,** 42–46 (1964)
55. Macdonald, D. R., and Webb, F. E. Insecticides and the spruce budworm. *Mem. Entomol. Soc. Can.*, **31,** 288–310 (1963)
56. MacPhee, A. W., and Sanford, K. H. The influence of spray programs on the fauna of apple orchards in Nova Scotia. VII. Effects on some beneficial arthropods. *Can. Entomologist*, **86,** 128–35 (1954)

57. Mattewman, W. G., Harcourt, G., and Perron, J. P. Timing of DDT applications for control of caterpillars on cabbage. *Can. Entomologist*, **84,** 346–52 (1952)
58. Mayer, J. Food and population: a different view. *Nutr. Rev.*, **22,** 353–57 (1964)
59. Mesnil, L. Nos connaissances actuelles sur les elaterides nuisibles en France. *Rev. Pathol. Vegetale Entomol. Agr. France*, **17,** 178–204 (1930)
60. Morris, R. F., Ed. The dynamics of epidemic spruce budworm populations. *Mem. Entomol. Soc. Can.*, **31,** 1–332 (1963)
61. Mules, J. H. W. Crutch strike by blowfly in sheep. A preventative operation. *Old. Agr. J.*, **44,** 237–41 (1935)
62. Mulla, M. S. An ecological basis for the suppression of *Hippelates* eye gnats. *J. Econ. Entomol.*, **56,** 678–70 (1963)
63. Newson, L. D., and Smith, C. E. Destruction of certain insect predators by applications of insecticides to control cotton pests. *J. Econ. Entomol.*, **42,** 904–8 (1949)
64. Nicholson, A. J. An outline of the dynamics of animal populations. *Australian J. Zool.*, **2,** 9–65 (1954)
65. Niklas, O. F. Die integrierte oder nützlingsschonende Schädlingsbekämpfung. *Allgem. Forstz.*, **19,** 415–16 (1964)
66. Painter, R. H. Resistance of plants to insects. *Ann. Rev. Entomol.*, **3,** 267–90 (1958)
67. Pence, R. J. The antimetabolite, imidazole as a pesticide. *J. Econ. Entomol.*, **56,** 1–7 (1963)
68. Pickett, A. D. The ecological effects of chemical control practices on arthropod populations in apple orchards in Nova Scotia. *Proc. Tech. Meeting Intern. Union for Conserv. of Nature and Natural Res., 8th, 1960, Warsaw*, 19–24 (1961)
69. Popham, W. L., and Hall, D. G. Insect eradication programs. *Ann. Rev. Entomol.*, **3,** 335–54 (1958)
70. Riordan, D. F. High-intensity flash discharge as a source of radiant energy for sterilizing insects. *Nature*, **204,** 1332 (1964)
71. Roach, W. A. Studies on possible causes of immunology. *Ann. Appl. Biol.*, **24,** 206–10 (1937)
72. Sayeed Quraishi, M., Osmani, M. H. and Hilal Ahmed, S. Effect of ultrasonic waves on the hatching of *Aedes aegypti* eggs at a frequency of 0.5 mc/s. *J. Econ. Entomol.*, **56,** 668–70 (1963)
73. Smith, C. N. Chemosterilants as a potential weapon for insect control. *Advan. Chem. Ser.*, **41,** 36–41 (1963)
74. Solomon, M. E. Dynamics of insect populations. *Ann. Rev. Entomol.*, **2,** 121–42 (1957)
75. Stern, V. M., Smith, R. F., van den Bosch, R., and Hagen, K. S. The integration of chemical and biological control of the spotted alfalfa aphid. The integrated control concept. *Hilgardia*, **29,** 81–101 (1959)
76. Tanada, Y. Microbial control of insect pests. *Ann. Rev. Entomol.*, **4,** 277–95 (1959)
77. Tansley, A. G. The use and abuse of vegetational concepts and terms. *Ecology*, **16,** 284–307 (1935)
78. Townsend, C. H. T. Methods of environment work for indicating insect control measures. *Ecology* **7,** 326–37 (1926)
79. Turnbull, A. L., and Chant, D. A. The practice and theory of biological control of insects in Canada. *Can. J. Zool.*, **39,** 697–753 (1961)
80. Vanderplank, F. L. Experiments in the hybridization of tsetse flies and the possibility of a new method of control. *Trans. Roy. Entomol. Soc. London*, **98,** 1–18 (1947)
81. Webb, F. E. Aerial forest spraying against spruce budworm—a problem of mutual interest in Canada and the United States. *J. Econ. Entomol.*, **53,** 631–33 (1960)
82. Westcott, C. *Halftruths or whole story, a review of Silent Spring* (Manufacturing Chemist's Assoc., Washington D. C., 4 pp., 1962)
83. Wigglesworth, V. B. The fauna of the orchard. *Rept. E. Malling Res. Sta. for 1959*, 39–43 (1960)
84. Wilkinson, P. R. Pasture spelling as a control measure for cattle ticks in southern Queensland. *Australian J. Agr. Res.*, **15,** 822–40 (1964)
85. Williams, L. L. Pesticides: A contribution to public health. *Am. J. Public Health*, **54,** No. 1, part 2, 32–36 (1964)
86. Woodworth, C. W. The theory of the parasitic control of insect pests. *Science*, **28,** 227–30 (1908)
87. Wright, D. P., Jr. Antifeeding compounds for insect control. *Advan. Chem. Ser.*, **41,** 56–63 (1963)
88. Wright, R. H. After pesticides—what? *Nature*, **204,** 121–25 (1964)

AN ECOLOGICAL APPROACH TO PEST CONTROL

by

P. W. GEIER AND L. R. CLARK

In common usage, the term "pest control" designates empirical attempts at curbing intolerably dense populations of noxious organisms.

The need for pest control, i.e. for relief from undesirable species, has increased to such an extent over the past half-century that a world-wide situation of unending struggle now exists. Since modern control began, we have become confronted by an increasing number of pest problems, the fundamental nature of which continues to be ignored in our attempts at solving them. An endless sequence of *ad hoc* programmes of destruction has kept the available trained personnel constantly mobilized in practical pursuits, with little time and energy for thinking about fundamentals.

This largely explains why the concepts which generally inspire pest control practices are so much out of step with modern ecological ideas. It must be conceded that the current approach to pest control has not progressed far beyond the original idea of striking down noxious populations by a series of blows delivered almost blindly but with a maximum of violence. This may be described as the "fly-swat approach".

We shall endeavour to indicate how ecological knowledge, and more particularly an understanding of population dynamics, ought to be utilized in pest control if lasting solutions are to be achieved. The basic ideas and descriptive terms employed in the discussion are those of NICHOLSON (1933, 1954, 1957, 1958). The essentials of his theory provide the most realistic basis available for a critical appreciation of the nature and problems of pest control.

The acceptance of such ideas leads us immediately to question the soundness of the term "pest control". A "pest" is a living organism which sometimes occurs in numbers inconvenient to man. The emphasis is on *population density*, i.e. on intensity of infestation, since it alone determines pest status, exclusively of any inherent property of the organism. In the field of crop production, one may usefully

IUNC SYMPOSIUM, Warsaw, 1960, vol. 7, pp. 15-24.

consider that pest status is attained only when a potentially injurious species reaches a threshold density at which it begins to cause a significant reduction in crop yield. However, for practical purposes, it is better to consider another more critical threshold density which may be defined as the level at which the loss caused by a pest just exceeds in value the cost of the control measures available. Here we have a relative criterion of injuriousness whose yardstick is fixed according to the general level of the economy at any time and place. In other words, an undesirable species is not an economic pest if it does not pay to combat it.

Thus, the rather ambiguous concept "pest" may be taken in two connotations—the first absolute—the second relative and economic. This distinction may be illustrated as follows. Assuming conditions of uniform infestation, kangaroos may reduce the carrying capacity of many Australian grazing properties for sheep, but it does not follow that kangaroo control would always be profitable. The only properties which would benefit by control are those where the increase in carrying capacity could be fully utilized to make up for the cost incurred. No immediate benefit would be derived from kangaroo control on other properties because a corresponding increase in financial return could not ensue. The difference between the two sorts of exploitation is one of economic intensity and technical development. In the first instance, the values of the absolute and economic threshold densities of the pest population are not very different. This is indicative of a high level of economic development. In the second instance, there is a large difference between the two threshold values—the density of the pest population would have to be substantially higher for control to be economical. Large differences between the absolute and economic thresholds occur at low levels of economic development. Since the value of the economic threshold density of a pest population will finally determine if and how protection is feasible commercially, no ecological assessment of a pest situation is complete without consideration of the economic and technical factors involved.

Since relative scarcity rather than great abundance is characteristic of most phytophagous organisms, one may wonder how species reach pest status. In most cases, if not all, they reach pest status as a consequence of man's activities. These activities are pest-inducing whenever they upset long established patterns of intra- and interspecific relationships, which have evolved in the course of natural selection. The disturbing influences of intensive land use, the breeding and distribution of man-improved organisms, the practice of monoculture, the introduction of exotic animals and plants, all contribute to the production of pest species, which, however, are surprisingly few in number.

Man's attempts to deal with his unintentional creation of pests has not only been inadequate; it has also been conducive to the production of further problems. In general, he has developed no better solution to pest problems than that of making his fly-swat bigger and more destructive. There is no essential difference between the effect of a fly-swat and that of, say, a modern insecticide applied indiscriminately. Each is a frontal attack on inconveniently dense populations delivered with the intention of knocking down numbers to a more acceptable level. Since eradication of the noxious species is not usually achieved by such means, their success is necessarily of short duration and, sooner or later, repeated application becomes indispensable. The fly-swat approach bears within itself the cause of its limited success—it can only reduce symptoms. It leaves unaffected the circumstances which induce pest-level densities.

This statement suggests that the term "control" must be divested of the meaning it assumes in the expression "pest control". There can in fact be no entirely man-made "control" of living organisms, since control is implicit in the existence of every persistent population. Every such population is kept under control as a matter of course by the interplay of its innate tendency to multiply and the inherent environmental opposition to population growth. This condition of continuing and universal control determines the state of balance in the absence of which populations would multiply indeterminately wherever individuals are able to reproduce at more than the replacement rate. Control, in its true meaning, bears no necessary relation to the anthropocentric concept of "pest". In general, pest populations are under natural control like any others, however inconvenient their numbers may be (Milne, 1959).

Since the density of a population is determined primarily by the intrinsic favourability of its environment, it follows that a population's vitality is not necessarily reduced in proportion to the numbers of individuals destroyed by an insecticide treatment or similar "control" measure. Thus, it is wrong to assume *a priori* that the greater the relative kill due to a treatment, the greater will be the long-term relief obtained. Populations of living organisms, as Nicholson has demonstrated by laboratory experimentation, are well-adapted to withstand high accidental mortalities. It appears that their resilience is a natural outcome of the governing mechanisms which implement natural control. Such mechanisms effectively oppose numerical increase when the population density is relatively high and facilitate it when numbers are relatively low. Any drastic reduction of population density due, for example, to the application of an insecticide, will induce the density governing mechanisms to relax the severity of their action to compensate for the losses incurred by the population. The relaxation induced may allow pest numbers to bounce back to peak population densities in as little as one generation. Compensation reactions probably explain the tendency for sprays to cause a sharp increase in the magnitude of the fluctuations normally displayed by pest populations.

In general, modification or elimination of the influence of density governing mechanisms probably accounts for the necessity of repeating treatments to keep numbers down in a population which has been reduced artificially. However, more may be involved than this: populations are genetically heterogeneous and often include small numbers of individuals "resistant" to the particular chemical agent used to limit population density. Relieved by insecticidal treatment from the competitive pressure of nonresistant forms, the resistant individuals may gradually form a new population against which the current means of protection are no longer effective. Moreover, many commonly-used pesticides are non-selective. Besides killing individuals of the pest population, they often affect numerous other species, harmless or even beneficial, with consequences similar to those described above. The void due to the treatment will soon be filled, under peculiar selective conditions which favour species with high reproductive capacities. Repeated treatment may result in a drastic change of faunal composition characterized by an overall reduction in the number of species present, and by a sharp increase in the abundance of surviving populations.

In brief, the fly-swat approach is unsatisfactory because it is applied without regard to the existing process of natural regulation to which all persistent populations of living organisms are subjected—however inadequate this limitation may appear to be for practical purposes. In blindly forcing alien mortality factors into balanced systems of natural control, current pest control practices not only fail to take advantage of the regulatory mechanisms available in nature, but even tend to interrupt or destroy the automatic operation of density governing agents.

Assuming these views to be correct, we obviously cannot solve the pest problem by relying only on natural control for protection against intolerable infestations. Having started to manage nature to our own ends, we can only overcome the unexpected difficulties encountered by the further development and extension of management. In other words, we must cease thinking in terms of traditional "pest control" and begin to think about pests in terms of *protective population management*. Considered in this light, protection against undesirable species appears as the mirror-image of species conservation, a field in which much lasting success has been achieved.

We must return to population theory to foresee how the protective management of noxious species can be achieved. The regulation of population density may be regarded as a dual process. Primarily, limits to numerical abundance are set "legislatively" by density factors (properties of essential requisites and properties of the species itself) which determine the intrinsic favourability of a given environment and the ability of a population to persist in it. The intrinsic favourability of a given environment may vary considerably from time to time. For an environment to be favourable to a species, it must at least generate a certain minimum supply of each requisite and provide a certain minimum level of security from attack or interference by inimical organisms. Otherwise the population could not persist.

In a favourable environment, the numbers of a species tend to increase progressively, eventually inducing an increase in environmental resistance by depleting one or more requisites, e.g. food, or by inducing an increase in the intensity of attack by natural enemies. The intrinsic environmental *rate of supply* of such a requisite, or the *level of security* from inimical organisms which the environment will permit, play a very important part in numerical regulation. However, these components of the favourable environment, being unaffected by population density, cannot adjust numbers to the prevailing conditions.

Numerical adjustment, i.e. density governance, which constitutes the second part of the dual process of regulation, is effected by the interaction of population density and the requisite brought into short supply by numerical increase. The governing agent is commonly a modification induced by population density in the availability of the requisite concerned, or in the intensity of attack by natural enemies: the mechanism of adjustment is usually some form of intraspecific competition. Where intensity of attack by a natural enemy is the environmental component involved in numerical adjustment, competition between individuals of the inimical population provides the mechanism. Density-induced processes hold populations about a constant or variable equilibrium density related to the properties of the species and to the rate of supply of the limiting requisite(s), both of which are modified by other components of the favourable environment, e.g. weather.

When faced with the problem of "managing" a population which has reached the status of an economic pest, an ecologist might decide either that the given environment favours, permanently or at irregular intervals, a general equilibrium level of population density which is intolerably high, or that the outbreaks of a pest are essentially the peaks of violent oscillations about an economically acceptable level. In the first case, management calls for a reduction in the general equilibrium level of the pest population, or stabilization at the low level which persists between outbreaks. In the second, management requires damping of the oscillations. In both cases, the approach would be either to induce a systematic reduction in the intrinsic favourability of the environment, or to modify adversely some specific properties of the noxious organism. In the present state of our knowledge, the former method seems the more promising generally. It could be implemented, for example, by reducing the supply of a requisite (e.g. shelter), or by increasing the overall intensity of attack by natural enemies.

The higher man's economic development of the environment, the greater are the possibilities for protective management. Where land use is primitive and agriculture extensive, protective management is necessarily reduced to its simplest form by lack of financial resources and limitations of technique. It can go no further, possibly, than the implantation of competitive or antagonistic organisms. A case in point is that of prickly pear and *Cactoblastis* (Nicholson, 1958, p. 109). As the level of economic development rises, so do the potentialities of protective management. The more man has altered the environment, the greater are the chances that some new or old environmental factor could be modified critically to the disadvantage of a pest without adversely affecting crop production. A perfect illustration is provided by the method of defence adopted against

Phylloxera vastatrix in European vineyards, where the economy of the wine industry permits the use of a range of vine rootstocks on which the insect cannot complete its life cycle.

Protective management does not imply the adoption or rejection of any mode of defence against undesirable populations, for it is not concerned primarily with means but with the way in which means are employed. For instance, there is certainly no incompatibility between protective management and the use of chemical pesticides. In fact, chemical pesticides represent an invaluable asset in protective management, initially as working tools and later as safety factors.

Where chemical pesticides have become indispensable, treatments obviously cannot be discontinued abruptly just because of their ecological inadequacies. The establishment of protective management must be gradual and may require a transitory period of considerable length before any significant departure from common routine has been achieved. Thus, in most cases, protective management must begin as an attempt to integrate current pest treatment with the recognized processes of natural control.

The first step in protective management is taken when treatments are no longer applied blindly, but made to conform objectively with local circumstances. This implies a systematic limitation of the disruptive interference with regulatory processes which a chemical treatment must entail. It calls for deliberate timing and careful selection of active ingredient, formulation, concentration, and method of application. Judicious timing requires accurate forecasting of pest population densities and a knowledge of the most vulnerable stage in the life cycle of an undesirable species. Moreover, timing should also restrict the effects of insecticidal treatments on harmless or beneficial populations. It is desirable that pesticides be not only effective but also selective in their action. The more specific a pesticide, the less likely it is to induce adverse side-effects. If eradication cannot be achieved, treatments should be designed to reduce noxious populations no more than is commercially desirable.

This first stage in protective management was reached fully—and profitably—several years ago in the apple orchards of Nova Scotia thanks to PICKETT and his co-workers (PICKETT *et al.*, 1946, 1953, 1958). Although crop protection is still achieved there largely by chemical means, the methods adopted have stabilized a pest situation previously as fluid and unpredictable as that faced by orchardists elsewhere.

In general, progress in the first stage will lead automatically to the second stage of protective management. This requires a working knowledge of the natural numerical regulation of the noxious species concerned, and, particularly, the identification of limiting density factors. It is effected directly by the deliberate modification of limiting density factors found to be susceptible to economical manipulation by man. Chemical destruction is by no means excluded at this stage, but treatments need no longer be as frequent or as heavy as before. Conversely, their efficacy is enhanced by diminution of the population pressure against which they have to measure. This stage, for example, has been attained successfully in the case of the cockchafer *Melolontha melolontha* in western Switzerland by MURBACH, BOURQUI & GEIER (1952).

The third possible stage in protective management is that of pest suppression as the ultimate consequence of comprehensive husbandry. This complete mastery is illustrated by the case of *Phylloxera vastatrix*. While this stage may sometimes be reached without detailed ecological study, such investigations will greatly enhance the chances of achieving it. The process of natural regulation is frequently both complex and subtle, involving critical density factors not readily identified by *ad hoc* methods.

The rate at which these successive stages may be reached depends firstly on the complexity of the initial situation, particularly on the number of potential pests to be dealt with simultaneously, and secondly on the intensity of the research effort devoted to the subject.

Nova Scotian experience suggests that the careful control of spraying is sufficient alone to allow the re-establishment of a measure of natural regulation which will maintain some undesirable populations below the economic threshold density (LORD, 1947, 1949). These populations are of species which attained economic pest status because of indiscriminate applications of pesticides. Their number may not be inconsiderable, and it is essential to distinguish them from the true problem species whose intolerable abundance is induced by the environment proper. Whereas the former need no longer be taken into account after Stage I, the latter must be dealt with by Stage II methods.

Whether Stage III can ever be reached for a majority of current pests is in fact not so much a matter of scientific possibility as of future economic expediency. It is economically expedient *now* to begin replacing pest control by some sort of protective management.

Although the changeover will require much creative effort from all concerned, it should prove to be less expensive in the long run than the continued use of present day methods.

Obviously, one cannot expect a proposal of this order to be implemented forthwith. However, we share with others the hope that the need for protective management will be realized before much more time and money have been wasted in the pursuit of remedies which ignore the ecological realities of pest control.

REFERENCES

LORD, F. T., 1947, The influence of spray programs on the fauna of apple orchards in Nova Scotia. II: Oystershell scale. Canad. Ent., **79**: 196-209.

—, 1949, The influence of spray programs on the fauna of apple orchards in Nova Scotia. III: Mites and their predators. Canad. Ent., **81**: 217-230.

MILNE, A., 1959, Weather, enemies and natural control of insect populations. J. Econ. Ent., **52**: 532-533.

MURBACH, R., BOURQUOI, P. & GEIER, P., 1952, Le hanneton et le ver blanc. Nouveaux enseignements, nouveaux procédés de lutte. Rev. Rom. Agric., **8**: 73-84.

NICHOLSON, A. J., 1933, The Balance of Animal Populations. J. Anim. Ecol., **2**: 132-178.

—, 1954, An outline of the dynamics of animal populations. Australian J. Zool., **2**: 9-65.

—, 1957, The self adjustment of populations to change. Cold Spr. Harb. Symp. Quant. Biol., **22**: 153-173.

—, 1958, Dynamics of insect populations. Ann. Rev. Ent., **3**: 107-136.

PICKETT, A. D., PATTERSON, N. A., STULTZ, H. T. & LORD, F. T., 1946, The influence of spray programs on the fauna of apple orchards in Nova Scotia, I: An appraisal of the problem and a method of approach. Sci. Agr., **26**: 590-600.

—, & PATTERSON N. A., 1953, The influence of spray programs on the fauna of apple orchards in Nova Scotia. IV: A review. Canad. Ent., **85**: 472-478.

—, PUTMAN, W. L. & LEROUX, E. J., 1958, Progress in harmonizing biological and chemical control of orchard pests in Eastern Canada. 10th Inter. Congr. Ent., **3**: 169-174.

ANALYSIS OF PEST SITUATIONS THROUGH THE LIFE SYSTEMS APPROACH

L. R. CLARK

THE ECOLOGICAL SYSTEMS APPROACH

The satisfactory management of arthropod populations is one of the many problems of human ecology and calls for a knowledge of those systems — ecosystems — which provide man with food, fibre, and other requirements.

The entomologist, whose task is to provide ways of managing the abundance and injuriousness of noxious arthropod populations, is most likely to succeed if he is familiar with the principles derived from the study of ecosystems, especially those pertaining to the fluctuation and regulation of population numbers. He should also possess a knowledge of the methods of population study and be provided with the material means to investigate pest situations adequately.

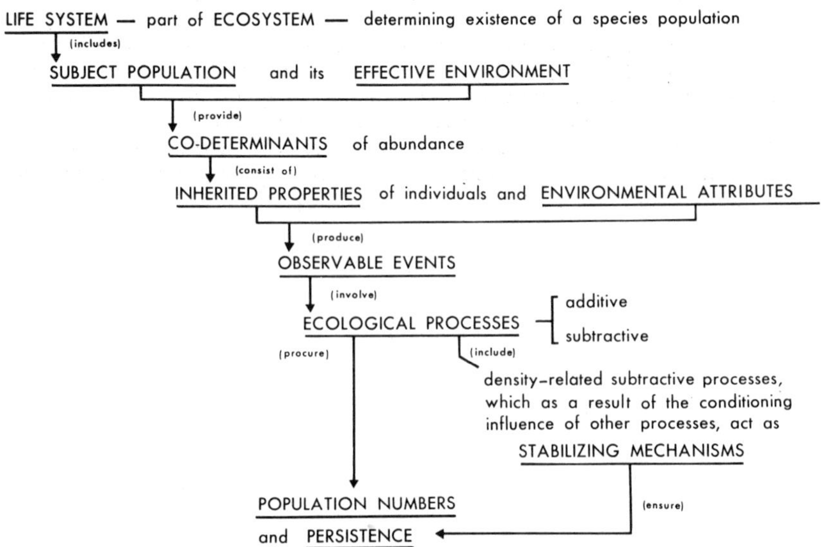

Figure 1. The elements involved in the functioning of a life system (after Clark et al.1967).

CONCEPTS OF PEST MANAGEMENT, Proceedings of Conference, March 25-27, 1970, pp. 45-57.

In our work on insect and other arthropod populations at Canberra, we have found it convenient to regard that part of an ecosystem which determines the abundance and evolution of a particular species population as a subsystem which we have termed the "life system" of that population (Clark et al.1967). We can define a life system as in Figure 1. We can think of the functioning of this kind of open-ended system as the operation of numerous ecological processes — some operating in sequence, some operating contemporaneously — some contributing towards the addition of individuals to the population, and others contributing towards the premature removal of individuals or to the limitation of natality. At least one of the ecological processes within a life system acts as a stabilizing mechanism, operating by negative feedback to limit numerical increase. In some systems, the operation of such density-related (density-dependent) processes exerts a major influence on the extent and form of numerical fluctations; in other systems, their influence in this respect is subordinate to that of density-independent processes such as the various actions of weather on population numbers.

The concept of life system is as flexible as the concept of ecosystem and the delimitation of a life system in space depends simply on the delimitation of the subject population. It is largely a matter of needs and convenience.

By definition, all of the many agents which act within a life system have some effect on the performance of the population. However, from the evidence obtained so far, it appears that only a few of those influences are largely responsible for the changes observed in population numbers, and for any changes that may occur in the quality of the individuals within the population. The majority of influences, some of which may be essential for the existence of the population, for example 0_2, have net effects which are almost constant or too small, relatively, to exert much influence on population trends (see Clark et al.1967).

This means that, if appropriate data are obtained on the major agents and processes involving them, there are reasonable prospects of producing a quantitative model to describe the functioning of a life system. Such models provide the best syntheses of life system studies because they show objectively how much is understood about a system. They can be refined from crude linear regression equations to constructs of much greater complexity and explanatory value, depending upon the requirements of a situation and the consequent amount of effort put into a study (see Watt 1968).

Developments in mathematical techniques and the availability of computers are making quantitative model building increasingly feasible and, during the next decade, we can expect to see the construction of quite a lot of life-system models for arthropod species. Some of the models for pest species will probably have sufficient predictive and explanatory value to be used in computer exercises to examine the likely outcome of various procedures of pest management, the most promising of which would be tested in the field. We may also see the construction of some "resource-centred" quantitative models of ecosystems incorporating sub-models which simulate the influence of pests.

UNDERSTANDING PEST SITUATIONS

We can define as pests those injurious or nuisance species the control of which is felt to be necessary either for economic or social reasons. We can define pest management as the satisfactory reduction of the effects of pests by any means applied with some ecological understanding. It is well known that arthropods such as insects and mites can achieve pest status in various ways, for example: by entry into the ecosystems of previously uncolonized regions; by adaptive changes in characteristics; by changes in human requirements or activities; and by increases in abundance as a result of environmental changes which are often caused by man.

Basically, two sorts of conditions prevail in the situations in which pest problems arise. One sort pertains to the characteristics of the pest species operating; to the characteristics of the resource or other human requirement affected; and to environmental influences such as climate. Such conditions are the outcome of evolution. For practical purposes, they are often treated quite reasonably as the "constants" of a pest situation, although it is realized that they may be undergoing forms of evolutionary change. The other conditions pertain to the functional aspects of the pest situation, that is to the actual interplay of the various influences involved *per medium* of ecological processes. Such conditions arise from the functioning of life systems, and determine the seriousness of a pest problem.

At present, man cannot readily modify the evolutionary facts of most pest situations, but he can influence ecological events and processes, so exploiting potentially favourable relationships. And, in fact, he has been endeavouring to do so for a long time with much success. That success, which has been achieved largely by an empirical approach, is not unqualified. Sometimes, it has been only temporary: in other instances it has been associated with the creation of further pest problems. In such situations, more knowledge is needed about the life systems involved.

It follows from considering the conditions under which pest situations arise that an attempt to understand and solve a pest problem should include three closely-linked phases of investigation which feed back upon one another (Figure 2).

PHASE I — DEALS WITH CAUSATION OF INJURY AND IDENTIFICATION AND DESCRIPTION OF THE MAJOR BIOLOGICAL AND PHYSICAL INFLUENCES INVOLVED

PHASE II — INVESTIGATION OF ECONOMICS OF INJURY AND FUNCTIONING OF THE RELEVANT LIFE SYSTEMS

PHASE III — MODIFICATION OF LIFE SYSTEMS TO ACHIEVE SATISFACTORY MANAGEMENT

Figure 2. The study of a pest situation.

Phase I:—The first phase, which deals essentially with qualities and qualitative relationships, has the object of establishing those facts in the pest situation which have been determined by evolution. The harmful species are identified and their characteristics are studied, for example, behaviour, powers of dispersal, reproductive ability, speed of development and adaptations to withstand adverse conditions. Consideration is given to the causation of injury and how the pest species interact. This entails study of the relevant characteristics of the resource or other requirement affected by the pests.

In addition, Phase I includes an investigation of the ecosystem or ecosystems in which the problem has arisen to identify the principal components of the effective environment of each pest population, including those which appear to favour the species — such as preferred or alternative sources of food, protective cover and favourable effects of weather — and those inimical to the species such as predators, pathogens and the adverse aspects of weather. Some knowledge of the characteristics and requirements of the natural enemies should also be acquired.

In other words, Phase I amounts to an intensive and systematic study of life histories, the object of which is to identify and describe the relevant elements of the pest situation, including the major components of the life system of each pest population.

Phase II: — The second phase, which deals essentially with quantities and quantitative relationships, is concerned with the actual operation of injurious interactions and the ecological processes entailed.

It includes, firstly, an assessment of the injuriousness of the pests. Injury and loss can be measured essentially in two ways: as the reduction in the quantity or quality of a resource or other requirement in terms of such biological values as biomass, productivity, yield, debilitation, or mortality; and, secondly, by expressing that reduction in monetary values. The latter is frequently a very difficult matter when many variables are involved in the interactions of supply and demand. However, some form of economic assessment is usually required to help decide the amount of injury or loss which can be tolerated, and how much effort should be devoted to quantitative ecological study.

The next step in Phase II is to proceed according to those decisions. There are four options:

- Firstly, it may be decided to take no further action, for example, because of the limited and sporadic occurrence of serious injury.
- Secondly, it may be judged that Phase I findings provide an adequate basis for decisions about management strategies and procedures. Usually this has been the decision made in the past.
- Thirdly, it may be decided to restrict quantitative study to some aspects of the functioning of the component life systems. This can range from monitoring pest numbers to intensive investigations of ecological processes undertaken, for example, to maximize the value of natural enemies, or to observe the outcome of minimizing some requirement of a pest species such as the amount of protective cover available.
- Fourthly, it may be judged that a more comprehensive approach is needed, and to undertake an analysis of the functioning of one or more life systems. If so, the task should be tackled with the idea of producing quantitative models to describe the determination of pest abundance and injuriousness.

Of the three options requiring action, only the last has any novelty as a possible approach for economic entomologists in general. It requires more thought and data than the empirical options, but it offers the prospect of providing the soundest possible basis for pest management and of yielding relevant facts and ideas which are likely to be missed by a quasi-intuitive approach.

Phase III: — The third phase of investigation completes the analysis of the pest situation by combining the findings of Phase I and Phase II. Decisions are made as to how the functioning of life systems could be usefully modified or circumvented and courses of action, or strategies, are adopted.

Four strategies are available for pest management (Clark et al 1967):

— The first is to *evade* the harmful effects of pests. One way is to make a resource inaccessible to pest attack. For example, seed potatoes are produced in areas which are almost free from virus-transmitting aphids. Alternatively, it is sometimes feasible to change the commercial form of a product to reduce the economic value of injuriousness, for example, by processing fruit or vegetables, including partly-damaged material which would otherwise be rejected.

— The second strategy is to *eliminate* the susceptibility of resource-producing plants and animals by making them resistant to pest attack in one way or another. A classical example is the elimination of the *Phylloxera* aphid from vineyards by the use of unfavourable root-stocks.

— The third strategy is to *suppress* those qualities of a pest which cause it to be injurious, for example, to remove infective sources from the life systems of disease vectors.

— The fourth is to *reduce* the numbers of pest populations to levels at which injuriousness is negligible. The extreme outcome of this strategy is, of course, extermination which has been achieved in a few instances, for example, screw worm in Curacao.

So far, relatively few pest problems have been solved by use of the first three strategies mentioned, but as biological knowledge increases the possibility of employing them more widely will also increase. The fourth strategy has been used more frequently and more successfully than the others. For this reason, and because it is likely to remain the principal course of action adopted in pest situations, it is worth considering in some detail.

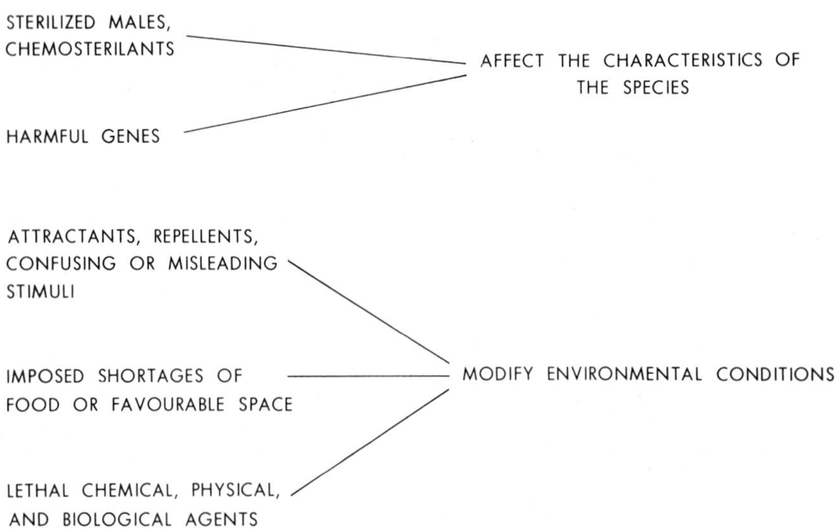

Figure 3. Possible means for limiting the numbers of a pest species (modified from Clark et al 1967).

REDUCING PEST NUMBERS

Attempts to limit the numbers of a pest species are made by introducing, into the life system of the population, one or more restraining agents with the aim of supplementing effectively the influence of those present naturally. Figure 3 lists restraining agents which can be used in pest management. The *intrinsic* effectiveness of those agents is measurable in terms of the long term protection provided and by the frequency with which human interventions are required to sustain their action (Clark et al.1967).

As already mentioned, man's ability to impose desirable modifications on the qualities of pest species is limited at present, as also is the extent to which he can interfere successfully with their normal behaviour. Much research is required before such means can be tested widely.

More success has been achieved with forms of environmental manipulation. In some situations, it has been possible to reduce the supply of one or more requisites needed for the existence of a pest species, for example, the supply of food or of protective cover, thereby adjusting population numbers satisfactorily in a lasting way. One of the more interesting current investigations known to me, in which this procedure is being tested, is the use of introduced dung-eating beetles to process cattle dung in northern Australia and thereby to reduce the numbers of the blood-sucking buffalo fly which breeds in the dung.

Most frequently, however, it has been necessary to rely on the introduction of agents which impose mortality directly on pest populations. In some life systems, these agents may cause density-related mortalities and assume a stabilizing role. In other systems, the proportion of individuals destroyed varies independently of the population density of the pest and population numbers are stabilized by interactions involving restraining agents present originally. Much success has been achieved by the use of chemical pesticides and, in numerous situations, satisfactory management has been achieved by the introduction of predators, parasites, and pathogens to life systems.

Because chemical pesticides are so widely used to limit pest numbers, it is desirable to concentrate our attention on management in situations in which they are the only restraining agent introduced. Chemicals are sometimes applied to check numerical increase whenever critical levels of abundance are reached. The pesticide is used as a stabilizing agent. Chemicals are also used in routine schedules applied during the same period each year. In the case of codling moth (*Carpocapsa pomonella* L.) in eastern Australia, the fraction of larvae destroyed varies independently of population density and the stabilization of population numbers depends on competition and cannibalism between the larvae which escape contact with pesticides and manage to enter fruits, and on competition for cocooning sites after feeding is completed (Geier 1964). In other instances, the routine applications of pesticides may tend to destroy a higher proportion of individuals when population numbers are relatively high than when they are low. Numbers may sometimes become high enough to induce significant competition between individuals for the food or living space which escapes contamination by the pesticide. Consequently, some individuals may be forced to leave uncontaminated places and be killed by the pesticide.

Whether they are applied in a density-independent manner or not, chemical pesticides have well-known disadvantages. They can certainly cause spectacular mortalities in particular generations of a pest population, but numbers in the following generation or two may be disappointingly high, for example, because of the favourable effects of weather on subsequent survival and reproduction or because of the destruction of natural enemies. Frequently application becomes an onerous task; sometimes undesirable side effects extend to the life systems of other species, and sooner or later pest populations develop resistance.

However, chemical pesticides are likely to remain a major means of pest management in the foreseeable future. It is necessary therefore to do all possible to minimize their disadvantages. This is partly a problem for economic entomologists and growers and partly one for the people who make and sell the chemicals, and for the governments of the countries in which they are made and used. It is a problem which calls for ecological understanding and a philosophy which, in principle, should regard the conventional application of chemical pesticides as a temporary means of management to be modified, supplemented by other means and, if feasible, replaced entirely by them.

In recent years, entomologists with ecological insight have made considerable progress in improving the application of chemical pesticides by using them as selectively and infrequently as possible to maximize the influence of natural restraining agents such as predators and parasites. In some instances, they have introduced natural enemies into the life systems of pests to supplement the use of pesticides or to replace them. In other cases, they have reduced the need for pesticides by modifying cultural practices.

This progress has been achieved mainly by commonsense, intuition, and a largely qualitative knowledge of pest populations and their environments. Further progress in depth calls for more quantification of life system studies and quantitative models to help optimize the use of pesticides.

This is the third time that I have referred to the desirability of constructing quantitative models of life systems for pest management. I know very well the large amount of effort that is required to build even a crude and simple model because several of my colleagues at Canberra and I are engaged in attempts at modelling. Such studies are costly in terms of manpower, money, and time, but so also are the amounts of *ad hoc* effort which have been devoted to the chemical control of some major pests over the years.

The question is — when is quantitative modelling justified in attempts at pest management by any of the available means? The practical answer is — when empirical attempts at management, guided by largely qualitative studies, do not yield a solution which is satisfactory both to the particular industry concerned and to the community in general, the latter being a matter for decision by a government or some funding body. The scientific answer is that the modelling approach seems to offer the only prospect of systematic progress in pest management, and therefore that technically-advanced, wealthy countries should at least set up some well-considered attempts at quantitative modelling to test the value of the approach. Some countries have already done so.

During the next 10 years we can expect to see a large expenditure of effort and money on attempts to develop relatively sophisticated means of control to replace conventional chemical pesticides, for example, chemosterilants, pheromones, juvenile hormones, and harmful genes. Such means are not likely to provide rapidly successful, universal remedies, and the people who test them in the field would benefit from having quantitative models of the life systems of target pests to provide guidance and to help in interpreting results. I think that the successful utilization of sophisticated means of control will require a sophisticated knowledge of the population biology of target pests. I say population biology rather than population ecology because we shall also need to take into consideration the genetics and adaptive powers of pest species (see Stehr 1969).

APPLE ORCHARD ECOSYSTEMS IN AUSTRALIA

To illustrate what I have said about the analysis of pest situations, let us consider the injurious arthropods of apple orchards in Australia. In 1957 we decided to begin intensive studies of those arthropods to increase our understanding of pest ecology with the idea of applying the knowledge gained in pest management. Apple orchards were chosen mainly for convenience of study and partly because some growers of pome fruits considered that the cost of chemical control was becoming prohibitive. The investigation was initiated by my colleague, P. W. Geier, who has remained the principal worker on apple orchard problems. For nine years he operated mainly in the Canberra district and in southern Victoria.

In 1966, the investigation was extended to New South Wales, South Australia, and Tasmania by the implementation of a Co-operative Research Programme supported partly by a growers' organization, the Apple and Pear Board, and partly by the Commonwealth Government. State entomologists and a senior State horticulturalist participated in the programme. They combined with CSIRO personnel to form an effective working party which was responsible for the collection and processing of data. The main object of the programme was to check and amplify the ecological knowledge gained during the first years of our work. Subsequently the CSIRO investigation was expanded in 1968 when J. L. Readshaw began a population study of orchard mites in eastern Australia and B. P. Springett began a similar study of San José scale (*Aspidiotus perniciosus* Comstock) in Western Australia where there is no codling moth problem. We now have a Joint Study Group composed of Geier, Readshaw, Springett, and myself, as chairman, to integrate our findings and future research efforts.

Our work on the arthropods of apple orchards began after codling moth had developed resistance to DDT (Smith 1955; Hooper 1968) and light brown apple moth (*Epiphyas postvittana* Walker), orchard mites (Tetranychidae) and wooly apple aphid (*Eriosoma lanigerum* Hausmann) had become serious pests, presumably because of the use of DDT. Soon after our work began, azinphosmethyl and carbaryl began to form the basis of spraying programmes (see Morris and van Baer 1959) although other OP compounds were also used extensively. So far, codling moth has not evolved resistance to such chemicals, but we have been less fortunate with other pests, for example, the light brown apple moth appears to have evolved strains resistant to azinphosethyl, carbaryl, and TDE. The two-spotted mite (*Tetranychus urticae* (Koch)) now has strains resistant to one or more of nine pesticides (R. W. Kerr 1969, personal communication).

The first candidate species for ecological studies was codling moth which, in mainland orchards, was found to be capable of destroying 85% or more of the crops on unsprayed trees every year. The favourable mainland climate permits the species to pass through three generation cycles annually. The adult moths which emerge in spring produce a first summer generation which is followed by a second and part of a third. A small percentage of surviving larvae of the first summer generation, most of the second, and all those of the third cycle enter diapause and form the overwintering component of populations.

Codling moth was important enough for both Phase I and Phase II investigations, the former being a intensive life history study (Geier 1963a,b) and the latter a population study which yielded a verbal model typical of the life system in unsprayed orchards near Canberra (Geier 1964). It was not possible to derive an objective monetary evaluation of injuriousness. However, that was unnecessary because of the obvious destructive ability of the species.

Briefly, the verbal model of the life system is as follows: the intrinsic reproductive ability of codling moth and the mainland climate make the species potentially capable of a 50-fold increase during the growing season of apples. However, increase to very high numbers is prevented by a double stabilizing mechanism: competition within fruits which entails some cannibalism; and competition for the cocooning sites in which the diapausing larvae overwinter. Of these two processes, the first serves to stabilize numerical trends within seasons, and the second adjusts numbers from one season to the next. Natural enemies are relatively unimportant because most of the mortality caused by them substitutes for mortality which would occur in the course of competition for fruit space and winter shelter (see Geier 1964).

The next candidate species selected was the indigenous light brown apple moth. It had become very injurious in Victorian and Tasmanian orchards, sometimes ruining whole crops, and little was known about its life systems. The results obtained by spraying with broad-spectrum pesticides were often disappointing and, as mentioned before, the species displayed considerable ability to develop resistance.

The Phase I investigation showed that light brown apple moth is a leaf roller which feeds on a wide range of dicotyledonous plants, including clovers and broad-leaved weeds which normally form part of the volunteer vegetation which colonizes apple orchards. Originally, light brown apple moth lived on acacias and other shrubs. In mainland orchards, it has three generations per year and overwinters as feeding larvae in the understory vegetation. Like those of codling moth, the adults are sedentary in habit, and the damaging infestations which occur in orchards are largely due to the build up of within-orchard populations. Light brown apple moth is favoured by climatic conditions which cause lush growth by its host plants. The larvae first feed mainly on the foliage of apple trees and later attack the fruits, damaging them superficially.

Phase II studies yielded life tables and convincing evidence that, in unsprayed orchards, natural enemies play a major part in the limitation of population numbers. They include spiders, bugs, several parasites, and an unidentified polyhedrosis virus. We cannot specify how each of the natural enemies operates and interacts with others until we have built a quantitative model. It seems, however, that the egg parasite, *Trichogramma* sp., acts in a density-related manner, and that spiders are probably the principal density-stabilizing agents. A recent discovery has delayed the attempt at modelling. It appears that, in some regions, large variations occur in the quality of populations from time to time, notably in the general vigour and reproductive ability of females. We need to be able to explain those differences.

We began the Cooperative Research Programme in 1966 armed with most of the information mentioned for codling moth and light brown apple moth, and Phase I type information for the other major pests. Some of that information was very scanty, at least for Australian conditions. The objectives of the programme, which utilized 8 orchards as study areas in 4 states, were to assess, at least in biological terms, the injuriousness of the orchard insects and mites; to learn how current spray programmes affected the injuriousness of the problem species; and to define optimal ways of minimizing their effects. As an experimental procedure, it was decided to use ryania 100 as a selective means of controlling codling moth and to act against the other pests only when they threatened to cause crop destruction or to kill trees. The abundance of the pest species and of some beneficial insects was monitored in appropriate ways ranging from population counts to rating procedures. The crops were harvested in autumn and examined for damage. Observations were made on the understory vegetation of the orchards, and records were kept of general orchard management activities, weather conditions, and irrigation.

During the 1966/67 season, ryania was applied in each study orchard according to an arbitrary schedule known to be effective for the Canberra area. The object was to gain some idea of how climatic and other differences between areas would affect the outcome of spraying. As expected, variable results were obtained in the percentages of fruits with deep entries by codling moth. It was found also that the appropriate spraying tactics would be to concentrate most effort on the larvae of the first summer generation, and to provide thereafter a minimal coverage to protect crops sufficiently.

With the help of a computer, our people then constructed some simple quantitative models. They used, in simplified form, what was known about the life systems of codling moth, the data on ryania applications, and information on the deterioration of spray deposits in the field. The models were built to determine objectively:

— the relative favourableness of each study area for codling moth;

— the actual effects of ryania on codling moth populations;

— the influence of local environmental conditions on the efficacy of ryania;

— and how to optimize ryania applications.

One of the models was used to work out spraying schedules for the following season, set to achieve 10-15 per cent of deep entries by codling moth in the harvested crops. Results of that order of magnitude were achieved for 5 of the 11 plots used, the percentages being much higher for the others. A single predictive construct was produced next, incorporating data for the 1967/68 season, and this was used to determine spray schedules to achieve 5 per cent deep entries in the harvested crops of 1968/69. For the seven study plots in which spray schedules were applied as planned, the percentages of deep entries were, in round figures, 1 (for trees with a very small crop), 3, 5, 5, 5, 8, and 9. This was naturally considered to be a satisfactory demonstration of the predictive value of the model. During the past few months, our people have refined the model further by incorporating data for 1968/69. We now have an adaptable construct which can be used for future simulation studies of possible control procedures anywhere in Australia.

The objective use of ryania sprays achieved satisfactory *demographic* control of codling moth but not satisfactory *economic* control. Even if we had budgetted successfully for only one per cent, deep entries in the harvested crops, *economic* control would not have been satisfactory. The reason is that ryania does not kill quickly enough to prevent a considerable percentage of poisoned larvae from penetrating the skins of fruits and causing unacceptable blemishes. That, however, was not important from our point of view because we were using ryania as a research tool and wanted to ascertain what would happen to the status of other pests when a selective pesticide was used against codling moth. The other pests responded in the following ways:

Light brown apple moth: In 1966/67 between 10 and 25 per cent of harvested fruits were damaged in three of the orchards; in the others the damage was negligible. The following season damage was serious only in one orchard, but in 1968/69 three crops were again severly damaged, one being from an orchard in which the injuriousness of light brown apple moth had been negligible previously. It was evident that, for this pest, natural control was inadequate to ensure economic control, at least in some areas, and hence that human interventions would be required from time to time to ensure crop protection. It is likely that, if a selective means of controlling codling moth could be developed commercially, the interventions would decrease with time because the species was not a major pest before the DDT era.

Two types of intervention were tested. The first was to destroy as much as possible of the understory vegetation by using a weed killer (diquat). This action caused a drastic reduction in the overwintering population. The other form of intervention was to apply pathogens as insecticidal sprays. *Bacillus thuringiensis* was tried first. The results were promising, with indications of a residual effect extending to the generation which developed after spraying. Subsequently, we tested the polyhedrosis mentioned earlier on a small number of trees, also with promising results.

Phytophagous mites: The numbers of both two-spotted mite and European red mite (*Panonychus ulmi* (Koch)) decreased substantially over the three-year period and, in 1968/69, no orchard plots were severely damaged.

The decrease in mite numbers was associated with increases in the numbers of predatory mites, chrysopids, and coccinellids, and appears to have been due largely to increased predation. This inference agrees with Readshaw's preliminary findings from his population studies of two-spotted mite and *Bryobia rubrioculus* (Scheut). His unpublished results suggest to us that a coccinellid (*Stethorus vagans* (Blackburn)) may play an important part in the numerical stabilization of two-spotted mite, and that a mirid (*Campylomma livida* Reuter) probably exerts a similar influence on *Bryobia*. Phytoseiids and chrysopids are also involved (Readshaw 1969, personal communication). It is beginning to look as though human interventions would rarely be necessary if broad spectrum pesticides were not used in apple orchards.

Woolly apple aphid: During the period of study, infestations were checked effectively in all instances by natural enemies including chrysopids and *Aphelinus mali* Hald., the former probably being the most important agents. It was inferred that, for woolly aphid, human interventions would be unnecessary in the absence of broad-spectrum pesticides.

San José scale: The results obtained from the programme for this species were less encouraging than for woolly apple aphid. Numbers of the scale increased markedly in the four study plots where it was recorded in 1966. At present, we do not know enough about the life systems of the species to interpret the population increases observed. From Springett's preliminary work in Western Australia, it appears that, in his unsprayed study orchard, the numbers of San José scale are stabilized at a very low level. However, the ecological processes involved have still to be identified. My guess is that a similar situation could develop elsewhere in the absence of broad-spectrum pesticides.

What do the cooperative studies mean for the future management of pests in our apple orchards? They indicate that, if an efficient selective means of control could be developed for codling moth, we would have much less trouble than at present with mites, woolly apple aphid, and probably light brown apple moth. San José scale would probably become less of a *potential* danger to trees than when broad-spectrum pesticides are used. In other words, granted a selective weapon for codling moth, there would be good prospects of managing the complex of pests by developing a programme of integrated control — integrated as between the life systems of the species concerned — and integrated as within life systems. By the latter, I mean well-informed interventions in the functioning of life systems when population numbers are likely to cause economic damage.

Recently, work began on the second stage of the Cooperative Research Programme. The following projects are involved:

— Victorian entomologists will evaluate the granulosis virus discovered by Falcon et al (1968) as a practical means for managing codling moth. This virus, or something like it, has been found to occur naturally in Victoria.
— Tasmanian entomologists will undertake further investigations on light brown apple moth, beginning with a study of population numbers in relation to damage, and experimenting probably with ways and means of reducing overwintering populations.
— South Australian workers will look into the possibility of controlling codling moth with reduced quantities of azinphosmethyl in an orchard which has not been sprayed for four years. The idea is to ascertain if codling moth can be controlled effectively by applications which do not interfere seriously with the predators of phytophagous mites.
— The New South Wales people plan to investigate the ecology and control of the plague thrips, (*Thrips imaginis* Bagnall). It is a sporadic pest of apple trees for which more information is needed.
— CSIRO will evaluate the polyhedrosis virus of light brown apple moth as a practical means of management. It will be tested both on apple trees and as a means of reducing the numbers of overwintering larvae.

We regard our research on orchard arthropods and the Cooperative Programme as preparation for the time when the present control procedure will cease to be feasible. There is no great urgency to produce alternative means of intervention because, for some years, we can expect the development and use of broad-spectrum pesticides to cope adequately in most instances with the pest situation. However, before very long, that control procedure will probably cease to be effective or acceptable for one or more reasons. For example: insoluble resistance problems may develop; the overall costs may become prohibitive; the health hazards involved may be considered too great for continuance; and conservationists may influence public opinion more than at present.

In ecological terms, we expect that the present control procedure will eventually cease to be effective because it interferes indiscriminately with the natural processes of ecosystems, with repercussions which are increasingly difficult to manage. Our aim is to understand those processes so that when the time comes for a change of approach we shall be able to recommend alternative procedures which exploit potentially favourable ecological relationships as much as possible.

ACKNOWLEDGMENT

Thanks are due to Methuen and Co., London, for permission to reproduce figures from the book "The Ecology of Insect Populations in Theory and Practice".

REFERENCES CITED

Clark, L. R., P. W. Geier, R. D. Hughes, and R. F. Morris. 1967. The Ecology of Insect Populations in Theory and Practice. Methuen and Co. Ltd., London.

Falcon, L. A., W. R. Kane, and R. S. Bethell. 1968. Preliminary evaluation of a granulosis virus for control of the codling moth. J. Econ. Entomol. 61:1208-1213.

Geier, P. W. 1963a. The life history of codling moth, *Cydia pomonella* (L.) (Lepidoptera: Tortricidae), in the Australian Capital Territory. Aust. J. Zool. 11:323-367.

Geier, P. W. 1963b. Wintering and spring emergence of codling moth, *Cydia pomonella* (L.) Lepidoptera: Tortricidae), in south-eastern Australia. Aust. J. Zool. 11:431-445.

Geier, P. W. 1964. Population dynamics of codling moth, *Cydia pomonella* (L.) (Tortricidae), in the Australian Capital Territory. Aust. J. Zool. 12:381-416.

Hooper, G. H. S. 1968. A review of the problem of insecticide resistance in Australia. J. Aust. Entomol. Soc. 7:67-76.

Kerr, R. W. 1969. Personal communication.

Morris, D. S., and R. van Baer. 1959. DDT resistant codling moth in Victoria. J. Dep. Agric. Vict. 57:619-623, 684-687.

Readshaw, J. L. 1969. Personal communication.

Smith, L. C. 1955. DDT resistant codling moth. J. Dep. Agric. S. Aust. 59: 12-15.

Stehr. G. 1969. On some concepts in the population biology of the spruce budworm. Proc. Entomol. Soc. Ontario 99:54-56.

Watt, K. E. F. 1968. Ecology and Resource Management. McGraw-Hill Book Co. Inc., New York.

NATURAL ENEMY COMPONENT

THE USE OF IMPORTED NATURAL ENEMIES IN INSECT PEST MANAGEMENT ECOLOGY

PAUL DeBACH

Introduction

NEARLY EVERYONE AGREES that biological control by importation of exotic natural enemies is a good idea. The main reason for this paper is that the idea isn't being adequately pursued. It may surprise some entomologists to know that many, if not most, of our serious pests have received little or no research work along this line. This, in spite of the fact that most major pests are invaders from abroad that presumably have left all or many of their natural enemies behind. I doubt that there is a single serious pest in this country which has had more than a small fraction of the potentially effective enemies discovered, imported and released against it. Many pest species have substantial lists of known enemies recorded, yet often minimal or no attempts have been made to import them.

I will attempt to analyze and clarify the relative importance and future potential in insect pest management ecology of the method of importation and colonization of new natural enemies of pest insects. I refer here to relative importance with respect to other research approaches or facets involved in pest management in agroecosystems. I feel strongly that we should stress research and allocate development funds on a priority basis toward the approaches most likely to solve pest management problems. This should be determined in relation to expectation of success from a given method as weighted against research and development cost and subsequent financial returns. Benefits are measured by reduction in damage and previous cost of treatment, increase in quality and decrease in environmental pollution or degradation. I believe we can set such priorities today by

TALL TIMBERS CONFERENCE ON ECOLOGICAL ANIMAL CONTROL BY HABITAT MANAGEMENT, 1971, pp. 211-233.

critical examinations of the past results achieved with various methods.

Lest I be misinterpreted, I do not advocate elimination or minimization of research on new ideas or pioneering-type approaches. I do suggest, however, that we not only be basic in our science but realistically businesslike as well. We do not have unlimited funds for research. Therefore, let's not spend $10,000,000 in research on, or pilot testing of, a new approach, such as (for example only) use of a pheromone or juvenile hormone, while we spend little or nothing on development of resistant plants or on the discovery of new natural enemies, especially when the latter already proven approaches may be investigated for comparatively little money.

I think we, as professional entomologists, should become even more scientifically critical of massive eradication projects on widespread pests by the use of pesticide chemicals. They are generally ecologically disruptive, have a high probability of failure and cost enormous amounts of money which might otherwise be used to better ends. Newsome (1965) states "at best the hope for eradication of established pests remains nearer a dream than a reality with currently available techniques." $200,000,000 is a figure I have read recently in connection with the proposed fire ant eradication program. It is tremendously exciting to think of how much good we could do in pest management ecology with that kind of money for research.

In proper pest management we stress ecological methods to maintain pest populations below economic damage levels with the minimum cost and with minimum adverse environmental effects. In a given crop or habitat, successful pest management may involve use of only one technique or type of manipulation by man or it may entail a combination of several. The latter will generally be the case, but in sugarcane in Hawaii completely successful management of all insect pests was achieved by importation of new natural enemies. On avocados in California virtually no pest management has been necessary because biological control occurs naturally. There is no magic formula that will give success just because we mix or integrate various techniques. Each case must be determined on its own bioecological merits.

Techniques in Pest Management

Ecological pest management generally will involve the active use of one or more manipulative methods or approaches by man against key pests, as well as his effort to conserve and optimize the degree of natural control of non-target insects which occurs unaided when not disrupted. Methods of manipulation proven in field usage include: (1) use of natural enemies; (2) application of cultural methods; (3) use of selective chemical pesticides; (4) development of resistant plants, and (5) use of irradiated sterile individuals. Other possible methods of manipulation which as yet remain largely unproven for practical pest management in agroecosystems include use of pheromones, electromagnetic energy, juvenile hormones, chemosterilants, repellants, antifeedants, and various genetic methods.

The status of the latter methods are summarized by various authors in Kilgore and Doutt (1967) essentially as follows: *Pheromones*—"to date no effective means of control of insect populations utilizing pheromones has been reported (page 256)"; *Electromagnetic energy*—only light traps are considered to show much promise in agriculture. Encouraging results have been obtained in some large scale experiments with the tomato and tobacco hornworms and others (page 124-128). Also, light traps are useful tools in sampling populations of various species; *Chemosterilants*—not yet developed as a commercial method (page 231); *Repellants*—no practical application now in agriculture (only method for facefly) (page 283); *Antifeedants*—"a growing amount of work has been done with . . . antifeedants but none so far have shown sufficient activity to be of commercial interest" (page 288). Likewise, genetic methods (aside from the use of irradiation-sterilized individuals) possibly may be utilized in a variety of ways, but currently even preliminary experiments to demonstrate control remain to be conducted (see Whitten, 1970).

Thus, for current use in agricultural pest management programs we are predominately dependent on four or five principal techniques or methods: biological control, selective chemical pesticides, cultural methods, selected resistant plants, and irradiation sterilized insects. In the excellent paper on Integrated Control presented by Smith and van den Bosch (*In:* Kilgore and Doutt, 1967), only three methods were stressed individually as being of broad significance in agricultural insect pest management—biological control, selective chemical pesticides and resistant plants. However, the importance of various cultural techniques in influencing pest populations both adversely and beneficially was implicit throughout their discussion.

What are the comparative number of successes that can be attributed to these methods, aside from chemical control? The advantages and disadvantages of the latter are well known and are capably summed up by Newsom (1965). However, our problem today is to reduce the use of chemicals while still maintaining satisfactory pest management. Thus I shall restrict comparisons to the aforementioned major non-chemical methods.

BIOLOGICAL CONTROL

Importation of new natural enemies alone on a world basis has resulted in the permanent control of 120 species. Forty-two of these pests were completely controlled; 48 species represent cases of substantial control and 30 of partial control (see Table 1 footnote for definitions of "complete" control, etc.). Additionally, various of these pest species have been subsequently controlled in other countries so that all told there have been 253 successful projects worldwide. This is by far the greatest number of successes achieved by any one non-chemical method, with the possible exception of cultural control. I have made no attempt to tabulate other cases of biological control arising from projects involving mass production and periodic colonization or other manipulations of established natural enemies, but a considerable number of practical field control projects of this nature are being carried on and the number is increasing yearly.

RESISTANT PLANT VARIETIES

According to Pathak (pp. 138-157 in Rabb and Guthrie, 1970), development of plant varieties resistant to the Hessian fly, wheat stem sawfly, and grape phylloxera are the main examples of this method. Newsome (1965) lists the wheat stem sawfly, Hessian fly, European corn borer, spotted alfalfa aphid, greenbug and chinch bug as being controlled more or less successfully on some crops in the United

States. I have not done a thorough literature research, so there are doubtless other cases, but the two reviewers mentioned probably covered the majority—some seven cases. I concur in Newsom's statement that it is surprising that so little attention has been devoted to this technique of pest control by entomologists. It has such obvious advantages it certainly should be exploited more thoroughly. However, he goes on to say that the method has some of the same limitations as chemical control—shifting physiological races of the pest pose the same problem to the plant breeder that insecticide resistance poses to the entomologist. Also, it may require an average of 10-15 years to develop a resistant variety. This means a relatively expensive and slow response to immediate problems.

IRRADIATION STERILIZED INSECTS

According to La Chance, Schmidt and Bushland (*In:* Kilgore and Doutt, 1967, pp. 148-193), successful demonstration of this method has been achieved with at least five insect species. They list four species as having been "eradicated" ranging from territories considered massive to small restricted localities. The successful eradication of the screwworm from Florida by 1959, and earlier from the island of Curaçao by 1955, remain the outstanding examples. A continuing massive program to eradicate the screwworm from the southwest United States and northern Mexico has been underway since 1962, during which time (1962-69) 43,348,738,774 sterile flies were released (Eddy and DeVaney, 1970). Although considerable control has occurred in Texas, eradication has not been attained. The cost of this program is several millions of dollars annually. The oriental fruit fly was eradicated from Guam in 1963 but tests on three other islands were not completely successful. The melon fly was eradicated from Rota Island in 1963 and the field cockchafer eradicated from a 74 acre plot in Switzerland in 1962.

In addition, tests with the Mediterranean fruit fly in Hawaii, with the Mexican fruit fly in Mexico, the Queensland fruit fly in Australia, and the codling moth in the United States, showed large enough reductions in crop damage to indicate this to be a promising method, either for control or perhaps for eradication under the right circumstances.

Some encouraging results or qualified successes are claimed for tests with the housefly and the navel orangeworm. Tests with the sheep blowfly failed to produce appreciable results.

Thus, about four species have been "eradicated," three of which occurred on relatively small "islands." Tests with another four species showed rather substantial degrees of control. To my knowledge, none of the latter cases have been adopted yet in regular pest management procedures. From these results, representing over a decade and a half of intensive and expensive research, it is obvious that this method, although spectacular and economically successful in specific cases, has limited application. This, of course, is not a very new or startling conclusion. According to Knipling (1964), "The sterile-insect technique will not be practical for controlling or eliminating established populations of most of our destructive insect species. The method will not be feasible for those that have a wide host range, are sporadic in appearance, or which do not cause high economic losses."

CULTURAL CONTROL

Cultural control techniques were more commonly used before the advent of organic insecticides made such methods seem cumbersome and complicated. Now that serious side problems have arisen from the use of insecticides, there is a strong resurgence of interest in cultural methods. Many have been and still are quite effective. However, there does not appear to be any broad review of the number, type and general effectiveness of cultural methods throughout the world so it is difficult to compare potentialities, costs and returns with other methods.

There is no doubt that such methods may be the key to achieving satisfactory pest management in a particular crop or area. Three research studies reported at the first Tall Timbers Conference illustrate this particularly well. Rabb's (1969) data on the tobacco hornworm clearly depict the importance of planting dates, early stalk destruction and sucker control in reducing pest populations. Van den Bosch and Stern (1969) showed that strikingly better pest management occurs if alfalfa is harvested in alternate strips on a chronological basis. Leigh, et al. (1969) showed that irrigation and fertilization regimes may have great impact on pest populations. These three studies also showed that the cultural methods used had a strong impact on the effectiveness of indigenous enemies. This indicates that in certain cases, perhaps many, the cultural technique operates indirectly by conserving or enhancing natural enemies, rather than by decimating pest populations directly.

Today, as a rough estimate, there may be from 100-200 cultural methods being used more or less effectively against a diversity of pests in various countries. According to my colleagues, H. T. Reynolds and V.M. Stern, probably a dozen are in use against cotton pests, including the pink bollworm, the boll weevil, the Egyptian cotton-leaf worm and the cotton-leaf perforator. Cultural methods may have good potential against most stalk boring insects and rootworms of annual crops where harvesting and plowing techniques may be manipulated to reduce pest populations. Also field breeding noxious Diptera can be controlled by use of the proper cultural practice that enhances intraspecific competition and natural enemy activity.

The practical application of scientifically sound cultural techniques may still remain difficult to achieve, and in some cases the program could be greatly weakened, unless nearly complete utilization throughout a wide area is adopted. Fairly expensive, in-depth, detailed and long-term ecological research must be contemplated for development of a cultural method because various alternatives and many interactions will be involved. As in the case of insecticides, it must be determined that the treatment to reduce the target pest does not have other effects which are adverse, such as upsets of non-target insect populations.

Of the four above-mentioned currently practical non-chemical methods available for increasing the extent of insect pest management, only biological control through importation of new natural enemies, and perhaps cultural control, have achieved a highly significant number of successes. Additionally, such biological control is inexpensive, it is permanent, and has no apparent drawbacks other than that final results may take from about 1 to 3 years following importation of a really effective natural enemy. However, this period applies only initially; thereafter the regulation achieved is *permanent*. Since we have been referring only to biological control from importation of natural enemies, this leads up to a consideration of the various other ways of manipulating natural enemies, as well as which of these may be the most advatageous under a given set of conditions.

Natural Enemy Manipulation

Natural enemies can be manipulated for use in pest management by (1) importation of new enemies from abroad, (2) conservation of already established enemies to improve or maintain their effectiveness through (a) provision of missing or inadequate requisites such as alternative hosts, supplementary food or shelter and (b) elimination or reduction of hazards or adverse environmental factors such as poor cultural practices, indiscriminate use of insecticides and other adverse physical or biotic factors, and (3) augmentation of already established natural enemies by periodic colonization or by genetic selection to improve fitness.

The priority importance of these three means of natural enemy manipulation in a given crop habitat must be determined by ecological studies of the faunal complex in chemically untreated plots as compared to treated ones in order to learn which of the pests present are "real" pests and which are man-made upsets resulting from pesticide chemicals or other cultural practices (see DeBach and Huffaker, 1971). Such studies constitute an absolutely essential first experimental procedure in development of any scientific pest management program. In addition to revealing actual pest status, they enable the determination of meaningful economic injury levels. If the pest(s) are real, in the sense that they lack effective established natural enemies, then importation of new natural enemies from abroad should be the primary major approach. If, on the other hand, the pest status is shown to be an induced or magnified one, i.e., (a) an upset of a non-target species or (b) a resurgence of a target pest, then in case (a) stress on means of conservation of natural enemies is indicated and in case (b) both improved conservation as well as concurrent importation of new natural enemies are indicated.

Should importation of new natural enemies fail to solve the problem of a real pest or a resurgent one after really adequate foreign search, then major emphasis should be switched to augmentation of natural enemies by periodic colonization or other means with continued minimum emphasis on foreign search. Thus it is clear that importation of new enemies from abroad should always be emphasized first and foremost in research with natural enemies, unless the pest in question is shown to be a secondarily upset species which already has *potentially* effective enemies established.

The foregoing should help to orient direction of research emphasis within the field of biological control endeavors. Because it indicates the overwhelming importance to pest management of importation of new natural enemies, a review of world successes in biological control importation projects seems apropos.

Review of Worldwide Results of Importation of Natural Enemies

The information and analysis that follow are based on published reports of essentially all projects that have been undertaken worldwide to achieve biological control through importation of new natural enemies. Summaries of these will be found in DeBach (1964) and Clausen (1971).

Natural enemies have been imported against some 223 species of insect pests and some degree of success has been obtained with 120 species, or over half of the species involved. The results are summarized in Table 1. The reader should note that first (Tables 1, 2 and 3) we are discussing only the number of *species* controlled. Later (Tables 4-7), inasmuch as many species have been controlled in more than one country, the total number of successful projects is discussed.

From Table 1 various conclusions and indications may be drawn. Completely successful control has been attained with 42 species, which means that these formerly serious and major pests no longer require chemical treatment in the country where the natural enemy was introduced. Ninety species are included in the combined categories of complete and substantial control. The latter category means that only occasional treatment is required, or perhaps that complete control was not achieved over the entire range of the pest. Successes have been achieved in nine orders but principally in the Diptera, Coleoptera, Lepidoptera and Homoptera. Over half of these oc-

TABLE 1. Number of pest insect species subjected to biological control attempts by the importation of natural enemies and the number and degree of successes attained. World total through 1969.

Order	*No. of pest species involved*	*No. and degree of successes** Partial	Substantial	Complete	Total	*No results*
Orthoptera	7	—	1	2	3	4
Dermaptera	1	—	1	—	1	—
Thysanoptera	3	—	—	—	—	3
Hymenoptera	9	1	2	1	4	5
Diptera	21	7	3	—	10	11
Coleoptera	37	7	4	2	13	24
Lepidoptera	51	6	14	4	24	27
Heteroptera	5	—	1	1	2	3
Homoptera	89	9	22	32	63	26
Total	223	30	48	42	120	103

* Complete successes refer to complete biological control being obtained and maintained against a major pest of a major crop over a fairly extensive area so that insecticidal treatment becomes rarely, if ever, necessary. Substantial successes will include cases where economic savings are somewhat less pronounced by reason of the pest or crop being less important, by the crop area being restricted (such as on a small island), or by the control being such that occasional insecticidal treatment is indicated. Partial successes are those where chemical control measures remain commonly necessary but either the intervals between necessary applications are lengthened or results are improved when the same treatments are used or outbreaks occur less frequently.

curred in the Homoptera. The reasons for this may be due to the fact that homopteran species are common invaders, were the first to be subjected to biological control, occur on expensive crops and traditionally have been difficult to control with insecticides. Thus a disproportionate amount of effort may have been directed against species of this order over the years. We have shown (DeBach, 1962) that success is more directly correlated with research emphasis than any other criterion. For those who might question the authenticity of the successes just referred to, it should be borne in mind that many (35 species) successes were replicated more than once (total, 135 repeat cases) both in time and space, so that the resultant statistical evidence gives a high level of proof.

In addition to the cases tabulated in Table 1, there are several unreported recent successes that I have observed personally or have been informed of in correspondence, as well as some that are probable successes but remain tentative because final results are not yet in. In Brazil the former two most serious scale insect pests of citrus (the Florida red scale, *Chrysomphalus ficus* Ashmead, and the purple scale, *Lepidosaphes beckii* (Newm.)) have been reduced to noneconomic rarities by the action of *Aphytis holoxanthus* DeBach against the former and of *A. lepidosaphes* Compere against the latter. I introduced these two parasite species into Brazil from California in 1962 under the auspices of the Rockefeller Foundation and the Brazilian government and observed the final successful results in 1970. In northern Peru the Florida red scale has likewise been reduced to an innocuous state by *Aphytis holoxanthus* which we sent at their request (personal observation). In Sicily the dictyosperum scale, *Chrysomphalus dictyospermi* (Morgan), and in Turkey the yellow and California red scales, *Aonidiella citrina* (Coquillett) and *A. aurantii* (Mask.) have been greatly reduced by *Aphytis melinus* imported from California (personal observation and correspondence.) In California the walnut aphid, *Chromaphis juglandicola* Kaltenback, appears to be in the final stages of complete control by a parasite, *Trioxys pallidus* Haliday, imported from Iran in 1968 (van den Bosch, et al. 1970). In Chile the woolly whitefly, *Aleurothrixus flocossus* (Maskell), is reported to have been completely controlled by the introduced parasite, *Amitus spiniferus* (Brethes) (personal correspondence), and in Tijuana, Mexico, and San Diego California, our studies indicate that this same woolly whitefly will soon be controlled by introduced parasites, principally *Amitus spiniferus* and *Eretmocerus paulistus* Hempel. Doubtless other current cases exist; those mentioned I happen to know of. These cases, and other recent ones in the literature, make it clear that biological control by importation of natural enemies has not reached a point of diminishing returns but rather needs increased emphasis. In fact many successes occur from accidental ecesis of natural enemies (DeBach, 1971). The purple scale is an outstanding example. It has been controlled in 9 countries by purposeful importation of its specific parasite, *Aphytis lepidosaphes* Compere, whereas it has been controlled in 13 countries by accidental transfer or immigration of *A. lepidosaphes*. Such fortuitous cases of biological control could all have been solved purposefully much earlier by man had the research effort been made.

Table 2 shows that the chances of achieving some significant degree of success against any given pest species from importation of

TABLE 2. Chances of success for controlling a given pest species by importation of natural enemies.*

Degree of biological control	*Pest species involved*	
	No.	*%*
Partial	30	13.5
Substantial	48	21.5 } 40.3
Complete	42	18.8 }
Subtotal	120	53.8
No results	103	46.2
Total	223	100

* See table 1 footnote for definitions of various degrees of biological control. Based on data in table 1, representing worldwide results. Includes only first project carried out against a given species. Subsequent transfer projects in other countries give an even greater probability of complete success (see table 5).

natural enemies are about 54 in 100. Substantial or complete success can be expected to occur with 4 pest species out of 10. This means that with about 40 percent of all species involved, a degree of control was attained that virtually eliminated the need for further insecticidal treatment. This is a very impressive success ratio. I do not intend to go into the economic achievements in this paper, but they are rather fantastic. The dollars returned in savings resulting from reduction in treatment costs and damage for each dollar invested has been projected as about 30 to 1, whereas for chemical pesticides it is about 5 to 1 (Pimentel, et al. 1965, pp. 227-291).

These results belie the opinions sometimes found in the literature to the effect that biological control is the best method of all when it works, but it is seldom successful. Thus it is clear that the following statement from the recent volume, "Insect-Pest Management and Control" (Subcommittee on Insect Pests, 1969), is misleading regarding the achievement of success—"The strategy has proved outstandingly successful in a substantial number of cases. *In many more cases, however, intensive and sustained efforts to apply biocontrol in this way failed utterly or, at best, produced only partial control, often of somewhat doubtful authenticity*" (author's italics). The unfortunate aspect of a statement like this is that it might inadvertantly reduce research emphasis when actually much more is justified and needed. What happens sometimes is that the non-specialist may may equate the proportion of imported natural enemy species that fail to become established with lack of project success in biological control. This is no more justified or realistic than evaluating the success of insecticides in pest control by comparing the number of original candidate chemicals with the number that become commercially used. This turns out to be a poor ratio of about 5,000 to 1 (Rümpker, Guest and Upholt, 1970) but, aside from their drawbacks, no one would deny that we have had many successful chemicals.

Candidate parasites and predators, i.e., the total number of entomophagous species imported for trial in biological control, are much more successful than this, and certainly much cheaper to test. Whitten (1970) estimates an establishment rate for natural enemies at between 20-30 percent and Carter (1970) gives figures indicating that about 22 percent of introduced natural enemies become established and of these about 17.4 percent were of "proved significance." However, in the long run, it is not the proportion established, or the proportion established to successes achieved, but rather the cost/benefit ratio that is really important and we have already shown this to be very good. The research and development cost of one new pesticide in the marketplace is now about $4,000,000 (Rümpker, Guest and Upholt, 1970). It is not necessarily a success even then; it still must prove itself in the field.

The research and development cost to obtain one new natural enemy will be a very small fraction of this. I have imported several into various countries with resulting impressive practical successes where the cost has been less than $100 per species. In other cases the cost may run much higher, but I believe not more than a few thousand dollars per entomophagous species at most.

The previous discussion has dealt with the chances of success in controlling a particular pest species and it was noted that different degrees of success might be obtained. Table 3 shows the proportion

TABLE 3. Relative chance of a particular degree of biological control being achieved against a given pest species.*

Degree of biological control	*Successes achieved*		
	No.	*%*	
Partial	30	25	
Substantial	48	40	75 (Substantial + Complete)
Complete	42	35	
Total	120	100	

* See table 1 footnote for definitions of various degrees of biological control. Based on data from table 1, but restricted to successful cases only. Includes only first project carried out against a given species. Subsequent transfer projects in other countries give an even greater probability of complete success (see table 5).

of the 3 categories of success to be expected on the average in cases where at least partial biological control occurs. It is seen that 75 percent of all successes are substantial or complete and it has already been pointed out that this means at the very least a great reduction in the previous use of pesticides. The 35 percent of successes in the "complete" category means, with rare or minor exceptions, that the pest problem is completely and permanently solved. Thus, for the most part, when some success is achieved in biological control it will be a significant one.

So far the analysis has been concerned only with the proportion of target pest species against which successful results have been achieved. However, successful results have been repeated with quite a few of these species in other countries. In the case of the cottony cushion scale, replicate successes have occurred in some 30 additional countries, and of the woolly apple aphid in 25. These represent the extra dividends in biological control. The original project may cost from a few hundred to many thousands of dollars, but the repeat or transfer projects often cost only the amount of an airmail shipment and some correspondence.

Table 4 includes the total of all successful projects from original importation as well as from subsequent transfer of natural enemies

TABLE 4. Total successes in biological control of insect pests by importation according to successful projects.* World total through 1969.

	Number and degree of success			
Order	*Partial*	*Substantial*	*Complete*	*Total*
Orthoptera	0	1	2	3
Dermaptera	1	1	0	2
Hymenoptera	1	3	1	5
Diptera	8	5	0	13
Coleoptera	7	7	2	16
Lepidoptera	9	24	4	37
Heteroptera	0	2	2	4
Homoptera	29	52	92	173
Total	55	95	103	253
%	21.7	37.6	40.7	100
		78.3		

* Includes original importation projects with a particular species as well as subsequent transfer or repeat projects in additional countries, i.e., the achievement of success against the 120 spp. included in table 1, led, in certain cases, to subsequent projects against the same pest species in other countries involving merely transfer of the already known and tested natural enemies. Thus first project successes against 120 pest species led to 253 total project successes.

to other countries. Instead of 120 successes if only species are tabulated, some 253 successful projects are seen to have resulted world-wide. Again, as in Table 3, it is clear that over ¾ of the successful projects are in the substantial or complete control category. The importance of these repeats should not be minimized. Biological control of a serious pest in a given country is just as important to them by transfer of already proven natural enemies as it would be had they been the first to solve the problem. If Florida entomologists should discover and import an effective parasite of the citrus snow scale, *Unaspis citri* (Comstock), this would be a major success and would go far toward making sound pest management possible. Its subsequent transfer would also benefit a substantial number of developing countries where the same scale is one of the most serious citrus pests.

The additional successes obtained by transfer of natural enemies is emphasized from another viewpoint in Table 5. This shows that

TABLE 5. Additional successes in biological control of insects derived from transfer of natural enemies from country of previous success to new countries.* World total through 1969.

	Additional number of successes achieved			
Order	*Partial*	*Substantial*	*Complete*	*Total*
Dermaptera	1	—	—	1
Hymenoptera	—	1	—	1
Diptera	1	2	—	3
Coleoptera	—	3	—	3
Lepidoptera	3	10	—	13
Heteroptera	—	1	1	2
Homoptera	20	30	60	110
Total	25	47	61	133
% increase over original results (see table 3)	83	99	145	111

* Difference between table 4 and table 1.

transfer projects not only more than doubled the original number of successes (133 added to the original 120), but that the increase in proportion of complete successes was considerably greater than in substantial successes, which in turn increased proportionately more than partial successes. Such results logically should be expected because the complete successes attract more notice, so are more likely to be repeated elsewhere.

Successes are tabulated in rank order in Table 6 according to country. The 10 leading countries have achieved over half of all the world

TABLE 6. Number of successful biological control importation projects by 20 leading "Countries."

Country	General climate	Number and degree of success			
		*P**	*S**	*C**	*Total*
1. Hawaii	Subtropical	4	10	9	23
2. U.S.A.	Temperate	8	10	4	22
3. California	Temp.-Subtrop.	7	8	6	21
4. Canada	Temperate	4	8	5	17
5. Australia	Temp.-Subtrop.	2	8	1	11
6. New Zealand	Temperate	1	6	3	10
7. Fiji	Tropical	3	1	3	7
8. Chile	Temp.-Subtrop.	3	1	3	7
9. South Africa	Subtropical	—	3	3	6
10. Peru	Subtropical	—	4	2	6
Sub Totals		32	59	39	130
11. Israel	Subtropical	—	1	4	5
12. Mauritius	Tropical	—	2	3	5
13. U.S.S.R.	Temperate	1	2	2	5
14. Tasmania	Temperate	1	4	—	5
15. Japan	Temp.-Subtrop.	—	—	4	4
16. Seychelles	Tropical	—	—	4	4
17. Mexico	Subtropical	—	2	2	4
18. Italy	Temp.-Subtrop.	1	1	2	4
19. Puerto Rico	Tropical	1	1	2	4
20. Greece	Temp.-Subtrop.	1	2	1	4
Sub Totals		5	15	24	44
21–66 All others		17	20	42	79
Grand Totals		54	94	105	253

* P = partial; S = substantial; C = complete (see table 1 footnote).

successes. It has been shown previously that these are the countries that have put the most effort into biological control research and importation (DeBach, 1962). In some cases one or a few individuals have been largely responsible for the progress.

Although salubrity of climate may have some influence on the probability of success in biological control, it certainly doesn't seem to be a generally serious limiting factor. When the 20 leading countries are grouped according to general climatic type, it is seen (Table 7) that the most successes occurred in temperate countries and the

TABLE 7. Biological control successes in 20 leading countries according to general climatic type.

Climate*	Number of countries	Degree of success**				% of successes in category	
		P	S	C	Total	S + C	C only
Tropical	4	4	4	12	20	80	60
Subtropical	5	4	20	20	44	91	45
Temperate-Subtropical	6	14	20	17	51	73	33
Temperate	5	15	30	14	59	75	24

* Tropical includes: Fiji, Mauritius, Seychelles, Puerto Rico
Subtropical includes: Hawaii, South Africa, Peru, Israel, Mexico
Temp.-Subtrop. includes: California, Australia, Chile, Japan, Italy, Greece
Temperature includes: U.S.A., Canada, New Zealand, U.S.S.R., Tasmania
** P = partial; S = substantial; C = complete (see table 1 footnote).

fewest in tropical ones. Again, this is interpreted as being related to the amount of research effort by the countries involved rather than to a correlation with climate. There seems, however, to be an indication that the proportion of complete successes may be greater in tropical areas. The important thing is that there doesn't appear to be any climatic zone where significant biological control of agricultural pests may not be achieved.

Guidelines for Importation of New Natural Enemies

It should go without saying that importation of new natural enemies should be in the hands of one or more properly trained specialists, with adequate quarantine facilities available. Ideally, one man should have overall responsibility for the entire project, including discovery and importation of new, carefully selected natural enemies, insectary quarantine screening and culture, field colonization, and evaluation of results.

If we are to emphasize importation of new natural enemies for pest management, as I am urging, certain principles and practices need elucidation.* This is especially necessary because some entomologists have made suggestions or raised questions that I consider erroneous or detrimental to the enhancement of biological control. Among them are the following ideas: (1) We should only import the single best natural enemy for a given pest, because interpecific competition with other less effective enemies may reduce total effectiveness; (2) We should not import any new natural enemies because the degree of biological control already achieved (even if not entirely satisfactory) may be lessened, i.e., competition between entomophaga may presumably lead to prey outbreaks; (3) We should not import, or at least not colonize, new natural enemies until preliminary ecological and biological studies of the existing faunal complex or of the new enemies have been completed—perhaps for 2 or 3 or more years. I do not concur with these ideas because they only limit the discovery, importation and colonization of scientifically selected and screened new natural enemies and thus delay or deny the possible achievement of biological control. In addition, such ideas have been completely worthless in themselves for predicting the outcome of any importation or for enhancing the achievement of success in any biological control project.

I do suggest the following in connection with importation policy and practice in biological control:

1. Exotic pests are obviously suitable subjects for biological control by importation of natural enemies, but native pests also have been controlled by imported enemies and should receive serious consideration. Effective natural enemies may be obtained abroad from either the target pest or from species related to it.

2. Direct pests (i.e., ones that directly damage the marketable item, such as fruit, at low population densities) are suitable objects for biological control but the probabilities for success may be some-

*This subject has been discussed previously in a paper presented at the 13th International Congress of Entomology, Moscow, August 1968, which is not yet published.

what lower than with indirect pests which can exist at higher population densities without causing economic damage.

3. Multiple importation of diverse natural enemies is the only practical manner of obtaining the best natural enemy for a given habitat, or the best combination of natural enemies for this habitat, or the best combination for the entire host range.

4. The competitive displacement principle reveals that competition between natural enemies normally is not detrimental to host population regulation; in fact, the displacement of an effective enemy species by another species means that the second is more effective and will produce even better host population regulation.

5. There usually is one best enemy for each pest species in a given habitat, and one may be sufficient for complete biological control. Often, however, a second or third enemy species may add to host population regulation and may in fact be necessary to achieve satisfactory biological control. The best enemy species may differ for different host habitats, hence there is generally no single best natural enemy throughout the range of a pest species. Therefore, all natural enemies of any promise should be imported and tried.

6. The best natural enemy may not be found until all natural enemies are known; hence, basic studies in biosystematics are of utmost importance and need increased emphasis and support. Currently knowledge is woefully inadequate.

7. From a *practical* standpoint, long-term basic ecological research on a pest species need not precede importation of new natural enemies. Such studies are not likely to help a *really effective* enemy and obviously the enemy cannot effect control unless it is imported. Any delay in discovery, importation and colonization of a really effective natural enemy merely lengthens the period that the species remains a pest.

8. The ultimate success of a given natural enemy candidate for importation *cannot be predicted in advance*, but we can find and select good risks. The effectiveness of natural enemies may be evaluated—experimentally or otherwise—in the country where they occur and their potential estimated. An effective natural enemy can be deduced to have the following characteristics: (a) high searching ability; (b) high degree of host specificity or preference; (c) good

reproductive capacity relative to the host, and (d) good adaptation to a wide range of environmental conditions. The most essential characteristic is high searching ability. It should be borne in mind that a really effective enemy may be scarce in its native home because it regulates the host population at low levels.

9. A good natural enemy becomes established easily and rapidly. Intensive and continued research effort should not be wasted on ones that don't. Rather, additional species or strains of species should be sought. Complicated and expensive programs are not necessarily required to achieve successful results.

10. Basic research on importation policy, and on the population ecology and genetics of colonizing species, should receive more emphasis but should not reduce in any way continued and increasing emphasis on current procedures for importation of new enemies.

11. No geographical area or crop or pest insect should be prejudged as being unsatisfactory for biological control attempts. The wide variety of successful results now obtained from importation of natural enemies suggests that nearly anything is possible.

Discussion and Conclusions

Hopefully I have shown that importation of new natural enemies from abroad should receive first priority in development of ecological pest management programs with much greater research emphasis and financial support being applied.

The method has good probability of success; it has general application; results are permanent; the cost-benefit ratio is very favorable with initial costs being low and recurrent costs nil; problems relative to insecticides are reduced or eliminated by its successful use, and no adverse effects on the ecosystem occur from biological control. No other method approaches this totality of favorable aspects.

Indications are great that many effective natural enemies await discovery, hence many opportunities are being neglected. Recent and past successes alone demonstrate this and many cases of accidental ecesis—distant dispersal of natural enemies resulting in fortuitous biological control unaided by man—support the idea.

As long as we ignore, anywhere in the world, an effective natural enemy capable of controlling one of our major pests, we are postponing cheap, reliable and permanent control until we import such enemies. The cost of other treatments and crop loss in the interim may justly be charged against the entomological profession.

A case in point is aptly illustrated by the statement of H. Compere (1969), "In 1879 Ashmead was attempting by correspondence to ascertain the native home of the Florida red scale, *Chrysomphalus ficus* Ash., to obtain its natural enemies for use in that state. *C. ficus* is reported to have been controlled in Florida recently by *Aphytis holoxanthus* DeBach. That should make an interesting story, since 85 years elapsed between the year that Ashmead attempted to ascertain the native home of *C. ficus* for the purpose of introducing its natural enemies and the introduction of the parasites that controlled it."

In other words, the Florida red scale remained a serious pest in Florida 85 years longer than it needed to. The same applies to certain other countries today that could repeat the Florida success merely by asking for a shipment of parasites. I think the same will be found to apply in many, many instances to a wide variety of pest insects.

Acknowledgments

This paper is a contribution of the IBP project on Biological Control of Scale Insects. Financial support was provided by NSF Grant GB-14489. My thanks are due to Dr. Fred Legner for critical review of the manuscript.

Literature Cited

Carter, Walter. 1970. The dynamics of entomology. Bull. Ent. Soc. America 16(4): 181–185.

Clausen, C. P. 1971. (Editor). Biological Control of Insect Pests and Weeds–A Review. United States Dept. Agr. Monograph., 1529 pp. (In Press.)

Compere, Harold. 1969. Changing trends and objectives in biological control. Proc. 1st Internat. Citrus Symp. 2: 755–764.

DeBach, Paul. 1962. An analysis of successes in biological control of insects in the Pacific area. Proc. Entomol. Soc. Hawaii. 18(1): 69–79.

————. 1964. (Editor). Biological Control of Insect Pests and Weeds. Reinhold Publ. Co., N.Y. 1–844.

————. 1971. Fortuitous biological control from ecesis of natural enemies. Kyushu Univ. Publns., Fukuoka, Japan. 30 pp. (In Press.)

DeBach, Paul and Carl B. Huffaker. 1971. Experimental techniques for evaluation of the effectiveness of natural enemies. In: Theory and Practice of Biological Control. Carl B. Huffaker, editor. Plenum Press, New York, 511 p.

Eddy, G. W., and J. A. DeVaney. 1970. A brief statistical review of the United States-Mexico screw-worm eradication program. Bull. Ent. Soc. America. 16(3): 159–164.

Kilgore, Wendell W., and Richard L. Doutt. 1967. Pest Control - Biological, Physical and Selected Chemical Methods. Academic Press, N.Y. 477 pp.

Knipling, E. F. 1964. The potential role of the sterility method for insect population control with special reference to combining this method with conventional methods. United States Dept. Agr., Agr. Res. Serv. 33–98.

Leigh, T. F., D. W. Grimes, H. Yamada, J. R. Stockton, and Dick Bassett. 1969. Arthropod abundance in cotton in relation to some cultural management variables. Proc. Tall Timbers Conf. Ecol. Animal Cont. Hab. Mang., 1: 71–83.

Newsom, L. D. 1965. Controlling Insect Pests and Diseases for Efficient Production for United States Agriculture. Proc. 14th Ann. Meet., Agr. Res. Instit., Natl. Acad. Sci., Washington, D.C.

Pimentel, David, et al. 1965. pp. 227–291. In: President's Science Advisory Committee 1965. Restoring the Quality of our Environment. The White House, Washington, D.C. 1–317.

Rabb, R. L. 1969. Environmental manipulations as influencing populations of tobacco hornworms. Proc. Tall Timbers Conf. Ecol. Animal Cont. Hab. Mang., 1: 175–191.

Rabb, R. L., and F. E. Guthrie. 1970. Concepts of pest management. North Carolina State Univ., Raleigh. 242 pp.

Subcommittee on Insect Pests. 1969. Insect-Pest Management and Control. Publ. 1965. Nat. Acad. Sci. 3: 1–508.

van den Bosch, Robert, and Vernon M. Stern. 1969. The effect of harvesting practices on insect populations in alfalfa. Proc. Tall Timbers Conf. Ecol. Animal Cont. Hab. Mang., 1: 47–54.

van den Bosch, Robert, B. D. Frazer, C. S. Davis, P. S. Messenger and Richard Hom. 1970. *Trioxys pallidus*, an effective new walnut aphid parasite from Iran. California Agric. 24(11): 8–10.

von Rümpker, R., H. R. Guest and W. M. Upholt. 1970. The search for safer, more selective and less persistent pesticides. Bioscience 20(18): 1004–1007.

Whitten, M. J. 1970. Genetics of pests in their management. 119–15. In: Rabb, R. L., and F. E. Guthrie, ed., 1970, Concepts of Pest Management, North Carolina State Univ., Raleigh. 1–242.

Descriptors: biological control; importation of natural enemies; pest management, techniques; pheromones; electromagnetic energy; chemosterilants; repellents; antifeedants; resistant plant varieties; irradiation sterilized insects; cultural control; scale insects; woolly whitefly.

INFLUENCE OF HOST DENSITY ON THE ABILITY OF SELECTIVE PARASITES TO MANAGE INSECT POPULATIONS

E. F. KNIPLING

THE INTERRELATIONSHIPS between host insects and the complex of parasites and predators that prey on them are probably less understood than those of any other animals. The number and diversity of the insect parasites and predators involved, the complexity of interacting factors, and the rapidity with which significant changes can take place make it difficult and costly to undertake more than a superficial study of predator-prey density relationships. However, even when studies have been undertaken in a natural environment, the information obtained has been more qualitative than quantitative. By and large, entomologists have not yet solved the important problem of determining the actual number of the host insect that is present at a given time and place, and even less information is available concerning the actual number of parasites and predators that prey on a given host species. With such a lack of quantitative information, it is understandable why so little is known about predator-prey density relationships, even in the case of some of our most destructive insect pests, those responsible for annual losses measured in terms of hundreds of millions of dollars.

Quantitative information about insect abundance, especially during the periods when populations are low, is recognized as being particularly important in the appraisal of the merits and limitations of control methods involving the use of genetic techniques or certain attractants. It is my firm belief that information about the numerical relationships between parasites and their hosts at various levels of host density is equally if not more important in an appraisal of the potentialities and limitations of insect parasites for the suppression or management of insect populations. In the absence of such quantitative information, it has been necessary to employ indirect ways of making estimates of population density for the purpose of appraising

TALL TIMBERS CONFERENCE ON ECOLOGICAL ANIMAL CONTROL BY HABITAT MANAGEMENT, 1970, pp. 3-21.

the feasibility of controlling insect species by such methods as the release of sterile insects, use of sex pheromones, or the liberation of parasites. Bits of information, the number of eggs recorded on 100 plants, the percentage of fruits attacked, the number of animals infested, and information about the biology and behavior of a pest are translated into estimates of the numbers of adults present per acre or per square mile or in a given region during the seasonal cycle. While such a method of estimating insect populations leaves much to be desired, it is encouraging that the estimates that were made for such insects as the screwworm, *Cochliomyia hominivorax* (Coquerel), the tobacco hornworm, *Protoparce sexta* (Johan.), and the boll weevil, *Anthonomus grandis* Bodeman, were reasonably close to more precise estimates made years later by more direct and more refined methods.

In recent years, I have become interested in estimating how many parasites, known to be important factors in natural control, have to be present naturally or would have to be released on a programmed basis to achieve practical control of some of our major insect pests, including two species of *Heliothis*, the sugarcane borer, *Diatraea saccharalis* (F.), and certain aphids. The release of parasites to achieve insect control has been considered for many years, but most such efforts have been unsuccessful or inconclusive. The failures may have resulted from a number of causes, but in my opinion, lack of understanding of the numerical relationships between parasites and hosts and testing on too small a scale were the chief causes of failure.

Entomologists have made substantial progress in the past 10 years in the mass colonization of a wide variety of insects. It seems within the realm of possibility to rear parasites and predators and make programmed releases of sufficient numbers to achieve practical control of some of our most destructive pests, especially species that now require costly applications of insecticides for control.

For effective use of programmed releases of parasites, it seems desirable to employ parasites that are reasonably selective against a given pest host. Selectivity for a given host or for a group of related hosts would seem to be desirable for two reasons. First, it seems logical to expect that a highly selective parasite will search for the desired host at all density levels. Second, a selective parasite should have no direct detrimental effect on beneficial insects or other organisms.

CLASSICAL CONCEPT OF THE QUANTITATIVE RELATIONSHIP OF A PARASITE AND ITS SELECTIVE HOST

Ecologists have recognized for many years that the size of a population of predaceous animals will fluctuate depending on the abundance of the prey species. In other words, the growth of the predaceous population must follow the growth of the population of its host(s) and collapse of the prey population leads to a collapse of the predaceous population. The same relationship exists between predaceous insects and their insect prey. However, a predaceous insect that feeds on a wide variety of insect hosts including other insect predators or even members of its own kind is not likely to be seriously affected, or at least will not be affected immediately, by fluctuations in the number of any given species of prey. In contrast, with a parasitic insect whose reproduction is restricted to a given host or to a group of related hosts, it becomes self evident that the density of the parasite population is governed by the density of the host population. In turn, the density of the host can be profoundly affected by the density of the parasite. Factors such as the searching efficiency and behavior, degree of mobility of the parasite or the host, kind and number of host plants or animals, climatic conditions, and many other factors can influence the parasite-host density relationship, but the dominating factor is the numerical relationship between the parasite and its selective host species.

One can postulate without risk of contradiction that the more selective a parasite becomes, the more precise and the more delicate will be the balance between the parasite and its selective host. Also, when a given parasite is dependent on a given host, one can make meaningful estimates of the number of parasites in the environment if information is available concerning the rate of parasitism and the actual number of host insects present in the same environment. For example, if it is established that the host density is 5,000 per acre during a given period and the rate of parasitism averages 25 percent, one can postulate that not more than 1,250 parasites will emerge from this number of hosts if each host normally produces one parasite. With populations throughout a large area, one can postulate that the maximum number of selective parasites generally will not approach the maximum number of the hosts in the total ecosystem.

The main point I wish to make is that if we had reasonably good estimates of the number of the host stage present and also had information about the levels of parasitism that occur at different levels of host density, we would be in a position to estimate the number of parasites needed to achieve various levels of parasitism. With such information we can, in turn, appraise the feasibility of rearing and releasing sufficient numbers of the parasite on a programmed basis to be useful in control. Eventually we might be in a position to prescribe methods of biological control to truly manage a highly destructive insect population throughout a large area. Therefore, information on the numerical relationship between parasites and their selective hosts correlated with information about the level of parasitism that results at various density levels will, in my opinion, be the key to the development of more effective and intelligent use of parasites for insect control in the future.

In the absence of information about the numerical relationship between parasites and their pest hosts and in view of the apparent difficulty in obtaining such information, theoretical studies were undertaken in an effort to estimate the limitations and potentialities of certain parasites as controls for some of our important insect pests. The studies thus far have been concentrated on the use of *Trichogramma* egg parasites to manage populations of bollworms (*Heliothis* spp.) on cotton and of sugarcane borers on sugarcane. However, similar studies have also been made of hymenopterous (*Apantales*) parasite for managing bollworms on cotton and aphids on potatoes and certain field crops. The results of the theoretical studies made to estimate the role of *Trichogramma* as parasites of bollworms on cotton and sugarcane borers on sugarcane were reported earlier (Knipling and McGuire 1966).

In the present paper, I will attempt to show by theoretical biomathematical analyses that the numerical relationship between a parasite and its selective host is a profound and dynamic phenomenon. Then suggestions will be offered for altering this relationship by manipulating the numbers of parasites or the numbers of hosts in the environment. *Trichogramma* will be considered first in this study and the host will be the sugarcane borer. We will then consider the role that selective parasites might play in managing *Heliothis* populations. The parameters for the studies on *Trichogramma* and the sugarcane borer will be similar to those previously established in the cited publication (Knipling and McGuire 1966).

THEORETICAL DENSITY RELATIONSHIP OF *TRICHOGRAMMA* AND THE SUGARCANE BORER

Trichogramma spp. are recognized as important parasites of the sugarcane borer, one of our most destructive insect pests. Thus, there are probably more data available concerning the degree of natural parasitism of sugarcane borer eggs by *Trichogramma* than concerning parasites by any other insect pest in this country. Such data have been obtained by State and Federal entomologists in Louisiana and by investigators in other countries where sugarcane is a major crop. Moreover, *Trichogramma* for some years were mass produced and released routinely in sugarcane fields to serve as an inoculum for natural control of the pest. Such releases were generally made sporadically at a rate of about 5,000 parasites per acre. Unfortunately, virtually all data obtained about rates of parasitism of sugarcane borer eggs by *Trichogramma* are qualitative. Little information is available about the number of parasites present or the number of host eggs present at the various levels of parasitism that were recorded.

In efforts to study the density relationship between a parasite and its host, it is also necessary to know how efficient the parasite is in finding its host in any given situation. There is virtually no information on this key point for field conditions. Consequently, in the theoretical studies of Knipling and McGuire (1966), values of efficiency were postulated to have a basis for making the calculations. It is not possible to say how accurate these postulated values are, but there is good reason to believe that they are sufficiently accurate to allow a good theoretical appraisal of the limitations and potentialities of the parasite for controlling such insects as bollworms on cotton and sugarcane borers on sugarcane. Recent results of investigations by F. D. Parker and others at the Biological Control of Insects Laboratory at Columbia, Missouri and by P. D. Lingren at the Cotton Insects Investigations Laboratory at Brownsville, Texas indicate that the estimated number of parasites required for effective control of the host are reasonably valid.

Various factors will influence the efficiency of a given number of parasites in host finding, among them the size of the host plant being searched for egg hosts. However, to simplify the models employed in this study, the basic assumption will be that the sugarcane being searched for eggs of the sugarcane borer is near maximum size. It is also assumed that 25,000 *Trichogramma* per acre searching during a period of 10 days, or the time required for one parasite generation, will be capable of finding and parasitizing 50 percent of the host eggs present in the area searched. With this as the basic efficiency value, more or fewer parasites are assumed to achieve levels of parasitism as follows:

Number of parasites per acre (10 days)	Percent parasitism of host eggs
1,000	2.8
5,000	13
10,000	24
15,000	34
20,000	42.6
25,000 assumed basic efficiency	*50*
50,000	75
75,000	87.5
100,000	93.8
125,000	96.9

The reasoning is as follows: If 25,000 parasites are capable of finding and parasitizing 50 percent of the host eggs in a given area and time, it is estimated that 25,000 additional parasites will find and parasitize 50 percent of the remaining eggs, which will increase parasitism to a level of 75 percent. By the same reasoning 75,000 parasites will achieve 87.5 percent parasitism. The larger the number of parasites, the longer the time spent in searching areas previously searched by other parasites; the fewer the number of parasites, the less the time spent in searching an area, and the more efficient each parasite becomes.

To estimate a numerical parasite-host egg relationship, we must also establish definitive numbers of the host that reflect the populations that exist in a natural environment. While great variations in the numbers of sugarcane borer eggs will occur from field to field, month to month, and year to year, enough information has been obtained from field observations to establish reasonably representative values in a seasonal trend. The following numbers are proposed as representative of an uncontrolled population of the sugarcane borer. One generation for the insect requires about 30 days.

Generation of host insect	Number of eggs per acre
1	10,000
2	50,000
3	150,000
4	300,000

A host egg density of 10,000 per acre each 10 days (30,000 per sugarcane borer generation) will be regarded as a low density population and is one that causes no significant economic damage in large sugarcane; 25,000 per acre each 10 days will be regarded as a moderately dense population and is one that will be regarded as economically damaging (economic threshold); and 100,000 per acre each 10 days will be regarded as a high density population and is one that will be regarded as highly damaging.

THEORETICAL EFFECT OF INOCULATIVE RELEASES OF *TRICHOGRAMMA* IN SUGARCANE FIELDS

As previously indicated, growers have employed *Trichogramma* releases in the past in hopes of controlling the sugarcane borer. Such releases generally involved inoculative releases of 5,000-10,000 parasites made with the expectation that the parasites would then increase to effective levels.

We will appraise the theoretical results of an inoculative release of 10,000 parasites during a 10-day period when the host egg densities are low, moderate, and high.

Low Host Egg Density

Host eggs per acre each 10 days	*Trichogramma* present	Percent parasitism	Number of eggs parasitized	Number of parasite progeny
10,000	10,000 (released)	24	2,400	2,400*
10,000	2,400 (progeny)	7	700	700
10,000	700 (progeny)	2	200	200

* Each parasitized egg generally produces 2 parasites. However, it is assumed that half the parasitized eggs will be lost because of predation and other causes so the number of parasite progeny will equal the number of eggs parasitized.

The model suggests that an inoculative release of 10,000 parasites made when the host density is low cannot result in an increase in the number of parasites. Indeed, the numbers of host eggs are far from adequate to maintain the original number of parasites released. The parasites would, therefore, rapidly decline in numbers and would be of decreasing significance in controlling the sugarcane borer.

Moderate Host Egg Density (Economic threshold)

Host eggs per acre each 10 days	Number *Trichogramma* present	Percent parasitism	Number of eggs parasitized	Number of parasite progeny
25,000	10,000 (released)	24	6,000	6,000
25,000	6,000 (natural)	15	3,750	3,750
25,000	3,750 (natural)	10	2,500	2,500

This model shows that even when the host egg density is moderate, the number of host eggs present is inadequate to maintain the original parasite population. The decline in number of parasites is less than for the previous model in which the host egg density is low, but the inoculative release would not result in the growth of an effective parasite population. The level of parasitism would be too low to provide much protection from damage by the sugarcane borer.

Host eggs per acre each 10 days	*High Host Egg Density* (Damaging level) Number *Trichogramma* present	Percent parasitism	Number of eggs parasitized	Number of parasite progeny
100,000	10,000 (released)	24	24,000	24,000
100,000	24,000 (progeny)	48	48,000	48,000
100,000	48,000 (progeny)	72	72,000	72,000

This model depicts the dramatic change that occurs in the dynamics of the parasite population at a high host egg density. The number of host eggs present is adequate to assure an increase in progeny. The increase in progeny in each generation, in turn, will lead to a higher and higher level of parasitism until the maximum potential is reached. Calculations are not shown, but if the sugarcane borer population remained constant at a level of 100,000 host eggs, the parasite population would continue to increase and would eventually stabilize at a level that would result in about 92 percent parasitism. Such a level of parasitism would theoretically reduce losses and would also cause a decline in the host population based on the growth rate of an uncontrolled population as shown on page 9.

The unfortunate feature of this model is that by the time the host eggs have reached a high density, a great deal of damage will have been done to the sugarcane. In other words, the three models suggest that a released population of *Trichogramma* is incapable of increasing naturally to effective levels before the pest host has substantially exceeded the economic threshold. This conclusion is not incompatible with the effects of natural parasitism observed in the field. Workers investigating the rate of parasitism report that high levels of parasitism do not generally occur naturally until late in the season when the population of sugarcane borer has already reached high and damaging levels.

While the models shown are hypothetical and the accuracy of the efficiency values for the parasite has not been fully determined, I am convinced that the estimates are sufficiently accurate to explain the failures of virtually all past efforts to control the sugarcane borer by making inoculative releases of *Trichogramma*. The natural laws governing the numerical relationship of the parasite and the host just will not permit this approach to provide an effective and dependable method of sugarcane borer control. In drawing this conclusion, I do not wish to imply that such releases would be of no value. Certainly, the high and increasing levels of parasitism that result following inoculative releases made when the host density is high would cause some reduction in economic losses. Also, from the standpoint of population dynamics, the high levels of parasitism shown, which should continue and actually increase during the fifth host generation, could be a significant factor in reducing the number of sugarcane borers that would develop and go into hibernation. This effect could be expected to reduce the next spring population and thereby lessen damage to young sugarcane and also delay the buildup of the pest to economically damaging levels until later in the season. However, such delayed control would not satisfy growers as a routine practice since they wish to reduce economic losses to a minimum. Moreover, parasitism often reaches high levels late in the season without inoculative releases of parasites.

Although I have concluded that *Trichogramma* egg parasites are incapable of achieving an effective level of natural control of the sugarcane borer by natural reproduction before the pest reaches the economic threshold and that similar limitations can be expected for *Trichogramma* and for certain larval parasites being considered for the control of *Heliothis* spp. on field crops, it would be unrealistic to conclude that the same limitations will apply to all parasites, all pest species, and all crops. Whether or not a selective parasite is capable of achieving economic control of a pest will depend on the pest density that causes unacceptable economic losses to a given crop and the density of the insect pest that becomes favorable for a high level of natural parasitism. It seems likely that pests which attack fruiting forms of a crop or which bore into stems are most likely to cause significant crop losses at densities that are too low for the buildup of parasites before substantial economic losses occur. On the other hand, certain crops are capable of tolerating insect pest densities that are sufficiently high to be favorable to the increase of selective parasites to effective levels before economic losses occur. Therefore, a better understanding of parasite-host density relationships correlated with information about the economic threshold of various pests will be important in determining the role that selective parasites can be expected to play as natural control agents. A better understanding of such interrelationships will also be needed to appraise the feasibility of using artificial means of increasing parasite populations to effective levels before economic losses occur.

PROGRAMMED RELEASES OF *TRICHOGRAMMA* FOR MANAGING SUGARCANE BORERS POPULATIONS

If *Trichogramma* cannot be expected to achieve satisfactory control of the sugarcane borer by natural increase resulting from inoculative releases, how can the parasite be used to assure satisfactory control of this pest? The answer is to release enough parasites on a sustained basis to achieve effective control. A hypothetical model will be established to show what might be expected from programmed releases in numbers that I feel are adequate to maintain control. It is assumed that the releases will begin when the natural population reaches the economic threshold of 25,000 host eggs per acre each 10 days. It is also assumed that 80 percent parasitism will be required to assure adequate control and also to keep the sugarcane borer population from increasing as the season advances.

According to the efficiency table shown on page 8, the release of 60,000 *Trichogramma* per acre over 10 days will achieve 80 percent parasitism. Thereafter, enough parasites will be added to supplement the normal production of progeny so as to maintain 80 percent parasitism. The requirements and results of such a program are shown in Table 1.

The program would theoretically maintain 80 percent parasitism by the release of a total of 260,000 parasites per acre during a period of two sugarcane borer generations. Such a hypothetical program, if realistic, would truly manage the insect population. An 80 percent level of parasitism at the assumed economic threshold of 25,000 eggs per acre each 10 days should provide assurance that no significant losses would occur. This level of parasitism should also keep the host population stable if the normal rate of increase for the insect is 5-fold.

If the programmed release of 260,000 parasites is a reasonably

Table 1. **A Hypothetical Program of Programmed Releases of *Trichogramma* for Managing a Sugarcane Borer Population**

Days	Host egg density	*Trichogramma* present Natural	Added	Total	Percent parasitism	No. parasite progeny
1–10	25,000	—	60,000	60,000	80	20,000
11–20	25,000	20,000	40,000	60,000	80	20,000
21–30	25,000	20,000	40,000	60,000	80	20,000
31–40	25,000	20,000	40,000	60,000	80	20,000
41–50	25,000	20,000	40,000	60,000	80	20,000
51–60	25,000	20,000	40,000	60,000	80	20,000

close estimate of the requirements for effective management of a sugarcane borer population after it reaches the economic threshold, it should eventually be practical to utilize such a method of control. The estimate of cost potentials for rearing *Trichogramma* made by the Columbia, Missouri Laboratory is less than 1¢ per 1,000 parasites. If we assume a production cost of $2.00 for 260,000 *Trichogramma* and a similar cost for application, the total cost would be $4.00 per acre. This would be less than the cost involved in the use of conventional chemical control. Moreover, such a program applied against the total population of sugarcane borers throughout a sugarcane growing area should reduce overwintered populations to the extent that the population would reach the economic threshold at least one generation later the next season. Thus, the necessary period of parasite releases would be reduced to one host generation instead of two, which could reduce annual costs well below the projected cost. However, reducing production costs is not the only goal of our research efforts. A major goal is to achieve insect control by methods that avoid environmental pollution.

ADDITION OF ALTERNATE HOST EGGS TO THE ENVIRONMENT

Since *Trichogramma* are obviously dependent on an adequate host egg density for the parasite to increase to effective levels of parasitism, I proposed some years ago that *Trichogramma* could be made more effective by adding sufficient numbers of suitable artificial host eggs to the environment. Such artificial host eggs might be an alternate host, sterile eggs deposited by released sterile females, sterile eggs of the host applied routinely, or, eventually, a truly artificial egg developed by insect nutritionists. Regardless of the source of such eggs, these theoretical projections suggest that it would be possible to completely reverse the normal host-egg parasite density relationship and in this manner maintain a parasite population at any desired or required level. For maintenance of 80 percent parasitism and stabilization of a sugarcane borer population at 25,000 host eggs per acre, the addition of 50,000 artificial host eggs per acre would be adequate, based on the various parameters established. The projected total of 75,000 host eggs per acre and a rate of parasitism of 80 percent would theoretically result in 60,000 parasitized host eggs and 60,000 parasite progeny each 10-day period.

The potential of such an approach to the suppression of an insect population will depend on advances in technology that must still be made. However, the method should provide an artificial means of maintaining a parasite density relationship that could never occur naturally in an environment.

RELEASE OF A SELECTIVE LARVAL PARASITE FOR MANAGING *HELIOTHIS* POPULATIONS

Preliminary appraisals have also been made to estimate the potential value of a selective larval parasite for suppressing populations of *Heliothis zea* (Boddie) and *H. virescens* (F.), two of our most destructive pests that now require the extensive use of insecticides for their control, especially on cotton and tobacco. Lingren at the Brownsville laboratory and W. J. Lewis at the Southern Grain Insects Investigations Laboratory at Tifton, Georgia have been investigating hymenopterous parasites which show a high degree of selectivity for *Heliothis* larvae. The information they have obtained concerning the natural incidence of parasitism has been of value in the effort to calculate the potential effectiveness of such parasites for managing *Heliothis* populations. Lingren (personal communications) reported that he observed natural parasitism of *Heliothis* spp. larvae by *Apanteles* spp. to reach a level of about 5 percent. I therefore postulated that during a period of 2 weeks 20 parasites per acre of cotton will be capable of finding and parasitizing 5 percent of the *Heliothis* larvae present during that time. Then in accordance with the method used in establishing the efficiency values given on page 8, I calculate that 280 parasites per acre during 2 weeks will be capable of finding and parasitizing 50 percent of the host larvae. Accordingly, 560 would be required for 75 percent parasitism, and 840 for 87.5 percent parasitism, etc. Although these assumed values of efficiency were originally postulated on the basis of very limited information, Lewis and Sparks (1970) have obtained information based on counts of field populations of *Cardiochiles nigreceps* (Viereck) and rates of parasitism in larvae of *H. virescens* which are of the same magnitude as those postulated.

Therefore, if we accept the efficiency values as postulated, we can, in turn, make an appraisal of the potential value of mass produced and programmed releases of a selective hymenopterous larval parasite for the management of *Heliothis* populations. Based on the progress that is being made in the mass production of parasites, especially at the Brownsville, Texas laboratory, it is estimated that in large-scale productions, the cost of rearing such parasites might eventually be as low as $2.00 per 1,000. We will appraise the effects of the release each 2 weeks of 200 parasites per acre of cotton. The assumed *Heliothis* population exists at an economic threshold of 5,000 larvae per acre each 2 weeks. The *Heliothis* population is increasing and, in the absence of control, would increase 3-fold from the assumed level of 5,000 larvae per acre to 15,000 per acre in the next generation. The theoretical effects of such a program are shown in Table 2.

The hypothetical program would effectively control the *Heliothis* population. The initial release of 200 parasites would achieve 40 percent parasitism, which should assure protection from crop losses when the host density exists at the economic threshold of 5,000 per acre during a 2-week period. However, this rate of induced parasitism will result in a substantial number of naturally produced parasites to supplement the programmed releases. This, in turn, will lead to an increasing rate of parasitism and even more progeny. Although,

Table 2. The calculated effect of the release of 200 hymenopterous parasites per acre each 2 weeks for the management of a *Heliothis* population on cotton.

Parasite generation	No. *Heliothis* larvae	Parasites Present			Percent parasitism	No. parasitized larvae	No. parasite progeny*
		Natural	Released	Total			
			Host generation 1				
1	5,000	0	200	200	40	2,000	200
2	5,000	200	200	400	60	3,000	300
			Host generation 2				
3	9,000	300	200	500	70	6,300	630
4	6,000	630	0	630	80	4,800	480

* Ten percent of parasitized larvae are assumed to produce adult parasites.

according to the parameters, the host population would increase slightly during the first half of the second generation, by the time this happened the number and rate of parasitism would be so high that no economic losses would be expected.

If the release of parasites as projected will, in fact, provide control of *Heliothis* on cotton, this could be an effective, selective, and low cost method of control. Only 600 parasites would have to be released to assure protection of the crop for 60 days. If parasite rearing costs are $1.20 per acre and release costs are 80¢ per acre additional, the total cost would be $2.00. Cotton growers often spend $10.00 or more per acre to control *Heliothis* on cotton with chemicals.

DISCUSSION

The accuracy of the numerical values obtained by these calculations may be questioned. Admittedly, the parameters used, the assumed host egg densities, larval densities, efficiency of the parasites, and number of parasite progeny produced, can only be regarded as rough estimates. Therefore, the projected effects could deviate substantially from actual results in a natural environment. However, there would seem to be little basis for questioning the validity of the principles governing the quantitative relationships of a selective parasite and its host, and I have considerable confidence that the projected requirements and projected results are reasonably accurate.

If the estimated numbers of *Trichogramma* parasites required to control the sugarcane borer are fairly close to actual requirements, this parasite could provide a practical alternative to chemical control. Likewise, the hymenopterous parasites of *Heliothis* spp. could provide practical control of these major pests. Other selective parasites of these and other lepidopterous insects may be equally or more effective. Much additional research on the mass rearing of both the host and candidate parasites and field evaluations of candidate parasites will be required before we can consider employing programmed releases of parasites for the control of some of our important lepidopterous insect pests. However, the results of this study, based largely on theoretical calculations, seem sufficiently promising to justify continued research to develop such approaches to insect control.

The initial objective might be the protection of limited acreages of crops by releases in community-wide programs, which may require substantial numbers of parasites to maintain high levels of parasitism. However, in managing total populations of some of our most destructive pests such as the *Heliothis* spp., we need not necessarily think in terms of maintaining high levels of parasitism. Perhaps an induced level of parasitism of the order of 30-40 percent that is reasonably constant in a large region would make the difference between noneconomic and economic populations throughout the region. Probably no insect species in all of its evolutionary history has ever been confronted with a moderately high and constant population of a selective natural enemy throughout its normal habitat regardless of the host density and with one which could almost immediately expand to higher levels with the expansion of the host population. We can only speculate about what the effects of such a phenomenon would be. However, there is every reason to believe that on a regional basis it would be a powerful suppression factor. For multiple generation species that often do not reach economically damaging levels until the third or fourth generation, the accumulative impact could be very great and could virtually assure noneconomic populations. For example, a suppression of 33⅓ percent above normal suppression each generation could theoretically reduce an insect population by more than two-thirds after three generations. Such a system of area suppression involving the release of a constant population of a selective parasite contains a "built-in" security factor that no other system of insect control possesses. If a host insect, because of abnormally favorable circumstances, should tend to increase to higher than normal numbers in spite of the release of parasites, the extra host resources should then automatically result in an increase in parasite progeny which, in turn, would result in a higher level of parasitism and greater suppression pressure on the host. An area-wide program to maintain constant low level suppression pressure, as described, probably would not endanger the existence of a key endemic species which many insect ecologists regard as important to assure a reasonable balance in the environment. What we might accomplish in such approach is to keep the pest reduced to a level representative of the normal environment before we added millions of acres of crops that permit the insect to increase to abnormally high levels.

Experiments to determine whether a regional approach to the management of insect populations as proposed is effective and economically feasible will be costly, and there are no financial resources for large-scale pilot test program of such an approach to insect population suppression. For example, information is needed on the number of parasites per unit area required to achieve different levels of parasitism, but valid information might not be obtained unless tests are conducted in a large area and on a sustained basis for two or more seasons. Pilot tests are necessary not only to determine the effectiveness of a method of population suppression but also to obtain cost data needed before we can estimate the economic feasibility of rearing billions of the parasites for release on a continuing basis in an entire region. However, when we consider the loss of millions of dollars caused by some of our major insects and the added millions spent on conventional insecticides for their control, we can hope that such tests might show that the cost of rearing the seemingly large numbers of parasites required would amount to only a fraction of the costs of control plus the annual losses caused by the pest.

As a long-range solution to the *Heliothis* problem, we might consider the suppression of the entire population east of the Mississippi. Perhaps the release of as few as 100 parasites per acre of susceptible host plants, both cultivated and wild, for a period of three generations starting about mid-May would assure the maintenance of noneconomic populations throughout the season. If this program required the rearing and release of several billion parasites at a cost of $10 million, the cost would probably be less than one-fourth the current cost of control with insecticides in the same area. In addition, such a procedure might virtually reduce all losses in such crops as cotton, tobacco, and soybeans that probably now amount to more than $100 million each year in spite of the costly investment in control by insecticides. Moreover, the method proposed should completely eliminate environmental pollution with insecticides, a matter of such grave concern at the present time.

I would like to offer one word of caution to those conducting research designed to test the principle of area suppression of insects. In view of the great mobility of many insects, any pilot experiments designed to suppress total populations by the release of parasites, sterile insects, or similar means need to be carefully planned and executed. Also, they must continue over a considerable period of time. Therefore, pilot-scale experiments on small islands or in ecologically isolated areas would seem to be necessary to insure a valid test of the principle of insect population suppression by such method as the programmed introduction of a selective parasite to the environment.

LITERATURE CITED

Knipling, E. F. and J. U. McGuire, Jr. 1966. Population models to appraise the limitation and potentialities of *Trichogramma* in managing host insect populations. USDA Tech. Bull. 1387, 44 p.

Lewis, W. J. and A. N. Sparks. 1970. Efficiency of *Cardiochiles nigriceps* Viereck as a parasite of *Heliothis virescens* on cotton. J. Econ. Entomol. (In press)

INVERTEBRATE PREDATORS

by R. I. SAILER

OF THE TWO WORDS that comprise my subject, each in a different way presents a problem. Zoologically the *invertebrates* include about every animal organism that cannot be classified as vertebrate. However, for purpose of discussion my comments will be restricted to arthropods and more specifically to the classes Insecta and Arachnida. The term *predator* is no less troublesome, for it is virtually impossible to accurately categorize the trophic behavior of many invertebrate species. Though it would be convenient to pigeonhole each species as phytophagous, parasitic, predacious, saprophytic, etc., this becomes a practical impossibility in the face of dietary habits that change from one life stage to another.

For example, what is one to do with a species such as *Itoplectis conquisitor* (Say). This hymenopterous insect develops as a pupal parasite of the gypsy moth. However, the adults may kill and feed on body fluids of as many as 250 pupae for each pupa successfully parasitized (*Campbell 1963*). When not stinging host pupae for feeding or oviposition, the adult female will feed, together with the adult males, on aphid honeydew and flower nectar.

The number of species exhibiting this type of parasitic behavior is very large. One family alone is estimated by Townes (*1970*) to contain 60,000 species. This would of course be the Ichneumonidae. Many years ago Wheeler (*1923*) coined the term *parasitoid* and applied it to these insects, which he regarded as not true parasites, but as extremely economical predators. Although in the past not generally accepted by entomologists, the term now seems likely to come into common usage. With

PROCEEDINGS OF THE 3RD ANNUAL N.E. FOREST INSECT WORK CONFERENCE, 1970.

the resurgent interest in biological control, there is need for a term that can be applied to the parasitic insects that will not convey the more restricted connotation and often opprobrious meaning of parasite.

However, the introduction and use of the term parasitoid points up my problem in discussing invertebrate predators. If I am to talk about the significance of predation by one insect on another, I can scarcely overlook the predatory activity of a very large number of parasitoid species.

Few people with any degree of familiarity with insects fail to be impressed by their remarkably diverse feeding habits. Certainly almost any conceivable dietary habit or combination of habits will be exhibited by one or more species of insect. However, I question whether even entomologists fully appreciate the extent to which the predacious habit pervades the insect world. If we were to list all orders, families, tribes, and genera of Insecta and Arachnida containing species that are predacious in some life stage, few groups above the tribal taxon would be omitted.

When we view invertebrates in terms of number of kinds and diversity of habit, it is obvious that a really conscientious treatment of invertebrate predators could profitably occupy much more than my allotted time. Therefore I propose to speak in generalities and to venture little more than some opinions regarding the role of predators in the population dynamics of prey species.

The Terms

So there be no misunderstanding about my use of terms when referring to different habits of insect groups, I will comment on distinctions between the terms *predator, parasitoid,* and *parasite*. Most people will agree that a predator is an organism that normally kills a more or less large number of serially contacted and consumed prey organisms. A parasite by contrast will live most of its life permanently attached to a single living host.

Unfortunately, the above definition of a parasite ill describes the behavior of most parasitic insects. Many species commonly referred to as parasites actually develop in hosts that are either killed by the parent parasite or die at some point in the development of the invading organism. Such parasitic insects are also usually free-living in the adult stage and in this stage may behave as predators.

Therefore, in further discussion of the importance of predators as compared with parasitoids and pathogens, remember that predacious activity of parasitoids will contribute to the functional importance of predation.

As indicated earlier, my principal objective is to consider the respective roles of predation, parasitism, and disease in regulation of host populations. Ultimately the question becomes one of knowing how these agents are to be manipulated for the purpose of holding pest populations at or below economic thresholds.

In 1929 W. R. Thompson wrote that though insect predators are numerous and are generally acknowledged to be beneficial, they are by no means universally believed to be as useful and important as insect parasites. Thompson then proceeded to present his reasons for believing that the part played by predacious insects was under-estimated. He argued that the relative importance of these two types of controlling agencies in any given case could be decided only by careful investigations in the field. Despite the recent life-table studies and mathematical models of R. F. Morris (*1963*), K. E. F. Watts (*1959, 1963*), C. S. Hollings (*1959, 1963*), and others it is my impression that our understanding of the relative role of predators in regulating pest populations has not advanced greatly beyond that of 1929. Perhaps it would be more correct to say that our ability to modify ecosystems through manipulation of predator populations has not advanced appreciably.

Though I make no pretense of being an active research worker in this field, I have for some 10 years now been involved in one way or another with biological-control programs. As a result of recent developments with which you are all well acquainted, biological control has suddenly moved from an almost stagnant backwater into the main current of research on pest control. Questions that could once be safely relegated to the ivory tower are now being hotly debated even in the popular press. Program-management people are finding such questions relevant and are pressing research personnel for answers that will not be immediately forthcoming.

Among the questions for which answers are needed, one of the most important is that of the relative values of the three major categories of natural enemies. If predators, parasitoids, and pathogens play different roles in regulating host populations, we must know what these roles are and how each can be used to the best advantage in the strategy of pest management. As a result of my own field experience and general familiarity with entomological research, I am convinced that invertebrate predators are very much more important than is generally realized. In fact I think there is good reason to believe that they constitute our first line of defense against the onslaught of insect pests.

The Value of Predators

Certainly we have examples that amply illustrate the potential value of predators. The first and most publicized example of biological control, the cottony-cushion scale, was accomplished by a predator—the vedalia beetle. Another coccinellid (*Cryptognatha nodiceps* Marshall) has controlled coconut scale (*Aspidiotus distructor* Signoret) in Fiji (*Taylor 1935*) and Principe (*Simmonds 1959*). A different kind of predator, the mirid bug *Cyrtorhinus mundulus* (Breddin), controlled the sugar cane leafhopper *Perkinsiella saccharacida* Kirkaldy in Hawaii (*Pemberton 1948*).

There are many well-documented cases of population release among previously innocuous plant-feeding species of insects and mites following use of organic insecticides. This is generally attributed to destruction of both predators and parasitoids. Evidence of this was reported by Hoffmann et al. (*1949*), who noted massive outbreaks of the mite *Paratetranychus ununguis* (Jac.) and several species of aphids in a Pennsylvania forest experimentally treated with DDT in 1945. They attributed these outbreaks to elimination of the abundant mite predator *Coniopteryx vicina* Hagen and drastic reduction of such aphid predators as *Anatis 15-punctata* (Olivier) and several species of syrphid flies.

Later examples of secondary pest problems that have followed in the wake of DDT and similar highly effective insecticides are too numerous and well known to warrant citing here. The likelihood that these problems have been precipitated in great part by the destruction of natural enemies is indicated by excellent experimental evidence. Perhaps the most convincing of the experimental studies are those of Franz (*1958*) and DeBach (*1955*). Franz was able to demonstrate highly significant increases of the balsam woolly aphid, *Adelges piceae* Ratz, in "plot" areas of tree trunk from which predators were excluded. DeBach used the insecticidal check method and concluded that DDT residues had no direct or indirect benefical effect on a diaspine scale population except through selective elimination of parasitoid *Aphytis chrysomphali* (Mercet). It is also well known that many scale insect and aphid outbreaks result from ants protecting them against predators and parasitoids (*DeBach, Dietrick and Fleschner 1951*).

From these various lines of evidence I believe most people are now willing to agree that natural enemy populations constitute a valuable stabilizing influence in ecosystems and are important factors in control of many pests. However, there is no general agreement about, or great interest in, the relative importance of predators, parasitoids, and pathogens as separate components of the natural enemy complex. Such comparisons as have been made have, for the most part, concerned the comparative ability of individual species to control pest populations. Any assessment of the value of predators as opposed to parasitoids or pathogens has consisted of little more than a tally of pests controlled by one as opposed to another. It is important that this question be investigated, for it may have great bearing on how natural enemies are managed.

Three Approaches

On the basis of present knowledge, there are three possible approaches to the use of predators and parasitoids for pest control:

1. The introduction of effective enemy species into niches where there are unexploited or inefficiently exploited food resources in the populations of pest species.

2. Modification of the environment to provide conditions favorable for increase of enemy species already present and able to exploit the pest populations.

3. The mass-production and inundative release of species capable of feeding on and destroying pest populations before they cause economic damage.

Of these approaches the first two impress me as the most practical and in the long run the least expensive. This is particularly true of pest problems associated with the forest habitat. Many of you know of the work that is being done in Europe and the USSR to utilize ants for control of certain forest pests. A summary of the European literature on this subject was published by G. Cotti in 1963. Other important references are one by K. Gösswald, G. Kneitz, and G. Schirmer in 1965 and an evaluation of the European research on use of red wood ants for protection of forests published by K. G. Adlung in 1966. That similar use of ants is being made in the USSR is indicated by papers published in the proceedings of a 1964 Symposium on Biological Control of Agricultural and Forest Pests edited by A. I. Cherepanov.

In Italy, under the direction of M. Pavan, extensive efforts have been made by the ministry of Agriculture and Forestry to relocate colonies of *Formica polyctena* Foerst in pine forests. As a result of a census conducted by 1,500 agents of the Italian State Forest Corps, Pavan (*1959*) calculated that in the Italian Alps there were more than 1,000,000 nests of four species of *Formica*. He estimated that these nests contained 300 billion workers that would destroy 14,400 tons of insects during an annual activity period of 200 days.

In Germany Gösswald (*1940*) has developed techniques for mass-breeding queens of *Formica rufa* F., and he has investigated the economic value of these ants. He has also actively promoted the colonization of ants in many forest situations for control of sawflies and lepidopterous pests. Ant hills are protected by law in Italy, Switzerland, and Germany; and I have been told by a German colleague that our military forces in Germany are billed for each ant hill destroyed in the course of maneuvers or other activities.

In Russia, Smirnov (*1964*) recommended that hills of the genus *Formica* be moved from forests that are being cut and placed in new plantations. Marikovskii (*1964*) noted that it was futile to move colonies to forests where competing species of ants were already well established. He urged that until more information was available and better success obtained from efforts to colonize the red wood ants, first priority should be placed on protection of existing colonies by means of State regulations and public education.

Adlung (*1966*) noted in his review clear evidence that ants are of value in controlling the common European forest pest *Panolis flammea* (Schiff.), and that the species *F. polyctena* has a certain limited value in pine forests. He concluded that establishment of nests is of no value in spruce plantations and would be injurious in beech forests, where the ants would protect and promote outbreaks of aphids that in turn cause the trees to be attacked by a canker fungus. He also noted that a considerable part of the ant's prey consists of beneficial insect species. Hence it may be concluded that ants are not an unmixed blessing even in situations where they do exercise effective control over pest populations.

An Appalling Task

There are of course many other examples of manipulation of predators, most involving movement of species from one country to another. In view of the number of pests that have invaded new continental and insular areas as well the certainty that predator species in many areas could be replaced by more efficient species resident in other areas, there is no doubt that much can be done to alleviate pest problems through introduction programs.

However, as a one-time taxonomist who has had some experience with insect ecology and its applied phase, biological control, I am appalled by the magnitude of the task that confronts entomologists if we seriously undertake to find and intelligently utilize all the invertebrate predators that are of potential value. There are thousands of species, a very large number of which are not even known to systematic entomologists. We need to know where to find these species and to know details of their life cycles, their ecological requirements, and what we might call their performance specifications. This boils down into a vast inventory job, supported by an information storage and retrieval system that would enable us to quickly determine what species of predator is available and capable of controlling a pest at a place where control is desired. In my opinion our present resources are totally inadequate to meet such a challenge.

The second approach is one that is almost equally formidable. It would require changes in agricultural practices and silviculture on a scale that would require many years if not decades to accomplish. At present we can only guess at what is possible. It is also questionable whether, in the face of the demands for food and space for our human population, the method would be economically feasible. Basically this approach constitutes the most important element of what is being referred to as integrated control or pest management. Although much talked about and experimentally demonstrated to be effective in California (*Stern et al. 1959*) the method has been purposely applied in only a few places and then on a limited scale. The most outstanding of these are applications in Nova Scotia against apple pests (*Pickett et al. 1958*) and in Peru against cotton pests (*Simon 1954*) and cocoa pests in Malaysia (*Conway 1969*). However, the dates of the first two suggests that there has been no great rush to put into practice what has been loudly preached in recent years.

The third method is one that offers promise of being effective in dealing with at least some pest problems. In the case of certain high economic return crops it may prove feasible to control pests by inundative or innoculative release of suitable predators. I am aware that most past efforts to use this technique have been unsuccessful. However I am encouraged by the results of a 1968 experiment conducted in Texas by R. L. Ridgway and S. E. Jones (*1969*). In this experiment two field releases totaling 292,000 *Chrysopa carnea* Stephens larvae per acre reduced the populations of *Heliothis zea* (Broddie) and *H. virescens* (F.) by as much as 96 percent and increased the yield of cotton in the test plots threefold. However, a word of caution is in order. The practical application of this method, even for control of *Heliothis* on cotton, will require extensive research on timing of release, numbers required, and most of all upon development of efficient and economical methods of mass-production and distribution of the predator.

Opportunities

I hope my comments will not be construed as presenting an unnecessarily pessimistic picture. Even with our present resources there are opportunities for significant progress toward solution of at least some of our pest problems through manipulation of predators. Furthermore, there is reason to believe that, within an ecosystem, predation acts as one of the most important factors in maintaining stability of the system. There are several lines of evidence suggesting that this is the case.

As a group, predators tend to be less specific in their food habits than parasitoids. They are less dependent upon a single food resource and are better able to maintain stable populations that tend to exploit prey species in proportion to their relative abundance. They also tend to be longer-lived and less likely to be adversely affected by physical factors of their environment. Their searching capabilities are generally greater, and in cases where a predator tends to be specific in its food habits there is likely to be a compensatory increase in either its fecundity or in its searching capacity. These are all characteristics that make predators well suited to play a specialized role in the regulative processes that restrain prey populations to levels below the carrying capacity of their habitat.

Nicholson (*1933*) Utida (*1955*) and others have experimentally confirmed the mathematically derived conclusion that interaction between a host and a parasite within a closed system results in population fluctuation of each species in oscillations of ever-increasing amplitudes to the point of mutual extinction. That such extinction rarely if ever occurs in nature is attributed by Voûte (*1951*) to the delayed-density-dependent action of predators. This action dampens the oscillations otherwise inherent in the host-parasite relationship. The direct effect is not so much on the prey population, but rather is due to intervention in the relationship between the prey and its parasites. The result is that the populations of both prey species and parasite are regulated at a more or less stable level. Voûte referred to this as "the quenching principle."

More recently Flanders (*1963*) has studied the interactions of food, prey, phytophage, predator, parasitoid, and pathogen components of an ecosystem. Within his system, Flanders found it necessary to introduce shelters that provided the phytophage a degree of protection "above the threshold of security" in order to maintain stability. This is clearly an example of Voûte's "quenching principle," and it introduces the element of space and shelter into population dynamics. However, of special interest to me was Flanders' evidence that disease epizootics occurred in his system followed elimination of the predator by periods of starvation due to host scarcity.

This sets the stage for what I suggest is a plausible and potentially useful classification of phytophagous arthropod populations and their associated natural enemies. In the tabulation that follows, populations are arranged in the order of four levels of abundance reflecting progressively increased density relative to carrying capacity of the habitat. The dominant regulative factor characteristic of each level is also shown:

Class	*Effect on ecosystem*	*Regulating factor*
1	Harmonic	Predators
2	Intrusive	Parasitoids
3	Disruptive	Pathogens
4	Catastrophic	Food

Though the above classification is grossly simplified, it is consistent with a very large body of empirical evidence derivable from the literature and experience of economic entomologists. However, examples will immediately come to mind that would seem to refute validity of such a classification. Perhaps the most striking is that of the Japanese beetle, *Popillia japonica*

Newman, a pest that is now generally believed to be regulated over large areas at a non-economic level by the pathogen *Bacillus popilliae* Dutky (*Fleming 1968*).

I regard the Japanese beetle as an exception that will tend to support the rule. Here the "quenching principle" function is provided by the characteristic habitat of the beetle larvae. The soil in which the larvae live provides a habitat in which the long lived *Bacillus* spores accumulate to a point where probability of non-infection is so high that few larvae survive to become adult. Obviously there are many other exceptions of a similar nature as well as examples that will doubtless exhibit almost every conceivable combination of two or more regulative factors working in conjunction. However, as a generalization that might serve as a useful guide for planning the strategy of biological control and pest management, such a classification could be most useful.

In any event I would urge that greater attention be given to the research on factors responsible for regulating non-economic populations. In the past entomologists have spent a disproportionate amount of time studying threatening or outbreak populations. We know little about what is happening to pest species during intervals between outbreaks or in areas where they are non-economic. If the role of invertebrate predators is a regulative factor responsible for restraining phytophagous insect populations to a harmonic part of their ecosystem, it would follow the biological control experts should place highest priority on the protection and augmentation of predacious species.

References

Adlung, K. G.
1966. A critical evaluation of the European research on use of red wood ants (Formica rufa group) for the protection of forests against harmful insects. Zeitschrift fur Angewandte Entomologie 57: 167-189.

Campbell, R. W.
1963. Some ichneumonid-sarcophagid interactions in the gypsy moth Porthetria dispar (L.) Lepidoptera: Lymantriidae. Can. Entomol. 95: 337-345.

Cherepanov, A. I. (Ed.)
1964. Biological control of agricultural and forest pests. Academy of Sciences USSR, Siberian Branch. Symp. 17-20 November, Proc. 1964: 179 pp. Novosbirsk. Translation by S. Nemchomok 1969 (available U.S. Dep. Commerce Clearing house for Federal and Technical Information Springfield, Virginia, 22151.)

Conway, Gordon R.
1969. A consequence of insecticides. Natural History, Special Suppl., Conf. Ecol. Aspects Int. Develop. Programs. Dec. 6-11, 1968. Warrenton, Virginia.

Cotti, G. 1963.
Bibliografia ragionata 1930-1961 del gruppo formica rufa in italiano, deutsch, english. Min. Agr. For. Collana Verde 8, Rome. 413 pp.

DeBach, Paul E. J. Dietrick and C. A. Fleschner.
1951. Ants versus biological control of citrus pests. Calif. Citograph 36: 312, 347-348.

DeBach, Paul.
1955. Validity of the insecticidal check method as a measure of the effectiveness of natural enemies of diaspine scale insects. J. Econ. Entomol.. 48: 584-588.

Flanders, S. E. and M. E. Badgley.
1963. Prey-predator interactions in self-balanced laboratory populations. Hilgardia 35(8): 143-183.

Fleming, W. E.
1968. Biological control of the Japanese beetle. USDA Tech. Bull. 1388, 78 pp.

Franz, J. M.
1958. The effectiveness of predators and food in limiting gradations of Adelges (Dreyfusia) piceae (Ratz.) in Europe. 10th Int. Congr. Entomol. Proc. (1956) 4: 781-787.

Gosswald, Von. K., G. Kneitz and G. Schirmer.
1965. Die geographische verbreitung der hugelbauden formica arten (Hy., Formicidae) in Europe. Zool. Jb. Syst. 92: 368-404.

Hoffman, C. H., H. L. Townes, H. H. Swift and R. I. Sailer.
1949. Field studies on the effects of airplane applications of DDT on forest invertebrates. Ecol. Monogr. 19: 1-46.

Holling, C. S.
1963. Some characteristics of simple types of predation and parasitism. Can. Entomol. 91: 385-398.

Holling, C. S.
1963. An experimental component analysis of population processes. Entomol. Soc. Can. Mem. 32: 22-32.

Huffaker, C. B.
1958. Experimental studies on predation: dispersion factors and predators-prey oscillations. Hilgardia 27: 343-383.

Marikovskii, P. I.
1964. Some problems involved in the utilization of ants for protection of forests against insect pests. See Cherepanov 1964, pages 129-131. loc. cit.

Morris, R. F.
1963. Predictive population equations based on key factors. Entomol. Soc. Can. Mem. 32: 16-22.

Nicholson, A. J.
1933. The balince of animal populations. J. Animal Ecol. (Suppl.) 2 (1): 132-178.

Pavan, M.
1959. Attivita italiana per la lotta biologica con formiche del gruppo formica rufa contro gli insetti dannosi alle foreste. Min. Agr. For. Collana Verde 4: 1-80. Rome.

Pemberton, C.E.
1948. HISTORY OF THE ENTOMOLOGY DEPARTMENT EXPERIMENT STATION, H.S.P.A. 1904-1945. Hawaiian Planters Rec. 52: 53-90.

Pickett, A. D., W. L. Rupman and E. J. LeRoux.
1958. PROGRESS IN HARMONIZING BIOLOGICA' AND CHEMICAL CONTROL OF ORCHARD PESTS IN EASTERN CANADA. 10th Int. Congr. Entomol. Proc. (1956) (3): 169-174.

Ridgeway, R. L. and S. E. Jones.
1969. INUNDATIVE RELEASES OF CHRYSOPA CARNEA FOR CONTROL OF HELIOTHIS ON COTTON J. Econ. Entomol. 62: 177-180.

Simon, F. J. E.
1954. SOME EXPERIMENTS ON CONTROL OF COTTON PESTS: INTERPLANTED CORN AND SYSTEMATIC INSECTICIDES. Inf. Estac, Exp. Agric. LaMolina No. 90. 22 p., illus. Lima.

Simmonds, F. J.
1959. BIOLOGICAL CONTROL—PAST, PRESENT, AND FUTURE. J. Econ. Entomol. 52: 1099-1102.

Smirnov, B. A.
1964. UTILIZATION OF ANTS IN PROTECTION OF FORESTS. see Cherepanov 1964, pages 152-154, loc. cit.

Stern, V. M., Smith, R. F., Van Den Bosch, R. and Hagen, K. S.
1959. THE INTEGRATION OF CHEMICAL AND BIOLOGICAL CONTROL OF THE SPOTTED ALFALFA APHID. PART I. THE INTEGRATED CONTROL CONCEPT. Hilgardia 29: 81-101.

Taylor, T. H. C.
1935. THE CAMPAIGN AGAINST ASPIDIOTUS DESTRUCTOR SIGN. IN FIJI. Entomol. Res. Bull. 26: 1-102.

Thompson, W. R.
1929. ON THE RELATIVE VALUE OF PARASITES AND PREDATORS IN THE BIOLOGICAL CONTROL OF INSECT PESTS. Entomol. Res. Bull. 19: 129-144.

Townes, H.
1969. THE GENERA OF ICHNEUMONIDAE. PART I. Amer. Entomol. Inst. Mem. 11, 300 pages. Ann Arbor, Michigan.

Voûte, A. D.
1951. ZUR FRAGE DEN REGULIERUNG DER INSEKTENPOPULATIONSDICHTE DURCH RAUBERISCHE TIERARTEN. Zeit. f. ang. Entomol. 33: 47-52.

Watt, K. E. F.
1959. A MATHEMATICAL MODEL OF THE EFFECT OF DENSITIES OF ATTACKING SPECIES ON THE NUMBER ATTACKED. Can. Entomol. 91: 129-144.

Watt, K. E. F.
1963. MATHEMATICAL POPULATION MODELS FOR FIVE AGRICULTURAL CROP PESTS. Entomol. Soc. Can. Mem. 32: 83-91.

Wheeler, W. M.
1923. SOCIAL LIFE AMONG THE INSECTS. 375 pp. Harcourt, Brace and Co. New York.

INSECT PATHOGENS AND THEIR POTENTIAL FOR REGULATION OF INSECT POPULATIONS

J. W. MacBAIN CAMERON

Like all other forms of animal life, insects are subject to diseases caused by infectious organisms. Some of these diseases have been recognized for well over 2000 years, and as might be guessed they were first observed and studied in useful species, namely, the honeybee and the silkworm, where they sometimes caused serious economic losses. As a matter of fact about 100 years ago the silk industry in France was on the verge of disaster due to the widespread occurrence of a disease caused by a protozoan.

Fungi

The disease-causing organisms are of four major kinds: bacteria, fungi, protozoa, and viruses. The fungi are probably the best known to the non-specialist, because infected insects often present a rather unusual appearance (Fig. 1). One of the most common of this group attacks the ordinary housefly, and everyone has probably seen the result—a rather mouldy-looking fly stuck tightly to window, and surrounded by a circular smudge on the glass. Grasshopper outbreaks have been frequently controlled by occurrence of a fungus disease. Tent caterpillars are sometimes the victim, also the hemlock looper. And many other species also from time to time are infected.

Despite their common occurrence, consistently successful management of entomogenous fungi has not yet been attained. First serious attempts to use them were made in Russia about 1884, but before any success had been achieved financial support for the plan was withdrawn. From time to time since then further tests have been undertaken, but without consistent success. This has usually been attrib-

TALL TIMBERS CONFERENCE ON ECOLOGICAL ANIMAL CONTROL BY HABITAT MANAGEMENT, 1971, pp. 267-277.

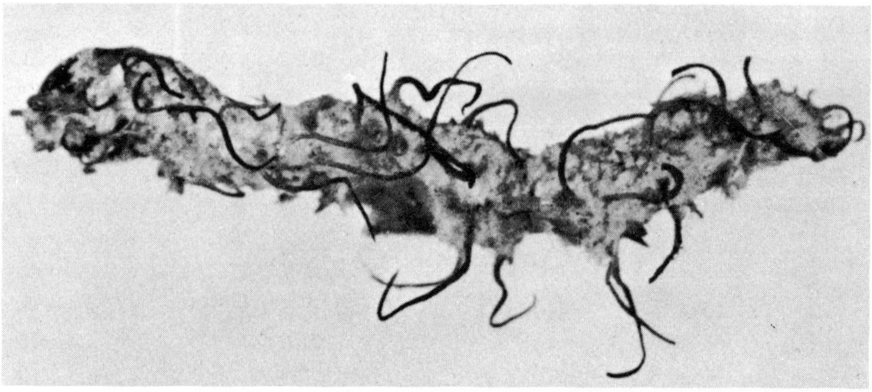

Fig. 1. Larva of spruce budworm, *Choristoneura fumiferana,* killed by a fungus belonging to the genus *Hirsutella.*

uted to unsatisfactory physical conditions, especially humidity, but there are probably other factors involved as well. Recently, and again first in Russia, combinations of fungi with very low dosages of chemical insecticides have been tried and promising results have been reported. However, at the present time there is no insect for which it is claimed that consistent control can be obtained by using fungi.

Protozoa

Similarly, the protozoa have not been found amenable to management, although as indicated above in the case of the silkworm they may at times cause serious epizootics. As records accumulate from the examination of insects in natural populations, it becomes more apparent that protozoa, particularly those belonging to the Class Microsporidia, are a very important ecological factor. In establishing laboratory rearings, we have had to be very careful to eliminate these organisms from wild parent stock by rigid selection. However, in spite of widespread natural occurrence they do not usually cause the collapse of an outbreak. Rather there appears to be a steady depressant effect, because infected individuals attain a smaller size and are of lower fecundity than are healthy ones. Sometimes this appears to be sufficient. There is one case reported from our Institute in which a microsporidian was credited with controlling an outbreak of spruce budworm after it had developed over a period of five years or so (Thomson, 1960). While admitting that this was a rather special case, owing to the isolated nature of the area, it nevertheless indicates that these pathogens do have a certain potential.

Most of the attempts to use protozoa have been made in Europe. Weiser (1957) has reported experiments in applying spores to trees infested by fall webworm with satisfactory control being achieved. A difficulty lies in that these are obligate parasites, and will grow and develop only in tissues of the host insect. This could be overcome by rearing and infecting the insects, but so far no method has been developed for storing the parasites to build up supplies or to carry them over winter. It therefore appears that considerable further research will be necessary before there can be any widespread application of protozoa as pathogens.

Bacteria

The great majority of bacteria commonly found associated with insects are harmless—a few are beneficial. But some, including some members of the so-called spore-formers, are the causal organisms of disease. Again study of the association between insect disease and bacteria covers nearly 100 years, and during the early part of the period claims were made that locusts and other serious pests could be controlled by spreading the infectious organisms. As in many other cases, disappointment followed. It was only after the Second World War that well-planned continuing research was undertaken, and as a result there are now a number of companies throughout the world, several of them in the United States, producing each year hundreds of tons of bacterial preparations for use in pest control.

The basis of the preparations is the bacterium *Bacillus thuringiensis*. There are at least ten or twelve recognized strains or varieties of this organism, but all have one characteristic in common—when the vegetative growth phase in artificial medium has been completed and sporulation is taking place, a proteinaceous parasporal body is formed that is toxic. All the susceptible insects are larvae of the Lepidoptera, and many of them are pests of greater or lesser importance in agriculture and forestry—cabbage looper, green cabbageworm, spruce budworm, hemlock looper, gypsy moth, tobacco hornworm, and others. To the present there is no evidence that this toxin affects beetles, flies or wasps. This is a fortunate coincidence, because while many pests belong to the Lepidoptera, most of the beneficial insects—parasites, predators and pollinators—belong to the other orders. Moreover, there has never been any evidence that this bacterium or its crystal toxin presents any danger to any animals other than the relatively few insects. This is recognized by the regulatory authorities, and the commercial preparations have been cleared for use on several food crops including cabbage, lettuce and cauliflower, without any tolerance limit being established. They are therefore very popular in truck crop areas because they can be used right up until harvest when neccessary. They can also be used effectively to protect shade trees from the attack of leaf-eating caterpillars (Fig. 2).

Field tests during the past two seasons have shown that these bacterial preparations can be used against the spruce budworm with

Fig. 2. Maple foliage protected from green-striped mapleworm, *Anisota rubicunda*, by a commercial preparation of *Bacillus thuringiensis*. Untreated on left.

some degree of success. Sprays can be applied at almost any time after the young larvae have emerged from the hibernacula and some mortality will occur. Counts made at the end of the season indicated reduction of the population by 50 to 95 percent, depending to some extent on the time of application, and the trees are visibly protected from defoliation. These tests were made using very heavy dosages, and much is still to be learned about the amount to use, the best time to apply it, and the best formulation. Research toward these goals is proceeding.

The bacterial insecticides are still relatively costly, and this will certainly limit their use for low value crops such as forest trees. But here too progress is being made. The search for new strains of greater potency is going on, and recently one of these has been formulated into a preparation that can be used for about a third the price that prevailed even a couple of years ago. This brings them within the range of cost for chemical insecticides. In those non-agricultural situations, such as perhaps park, resort or picnic areas, or valuable fishing streams or lakes, where usage is high and chemicals might be objectionable, they should be seriously considered. Also, bearing in mind that they do not affect parasites and predators, their use might result in a relative increase in these other regulating agents to a point where the natural balance would be restored at an acceptable level of the pest.

Viruses

Viruses present a still different picture. They are obligate parasites, and are normally produced only in the cells of their host. As a question in semantics it might be argued that they are not organisms. Essentially they consist of a core of nucleic acid enclosed in a protein coat, the whole being called a virion. The virions in turn are occluded, singly or in considerable numbers, in an inert protein inclusion body (Fig. 3). When ingested by a susceptible insect the occluding protein is digested or dissolved and the virion penetrates a cell where it assumes control of the intracellular metabolic processes and more virus is produced. Eventually the tissue is destroyed and the insect dies. This is a true infection, and the virus in the insect may be multiplied many thousandfold.

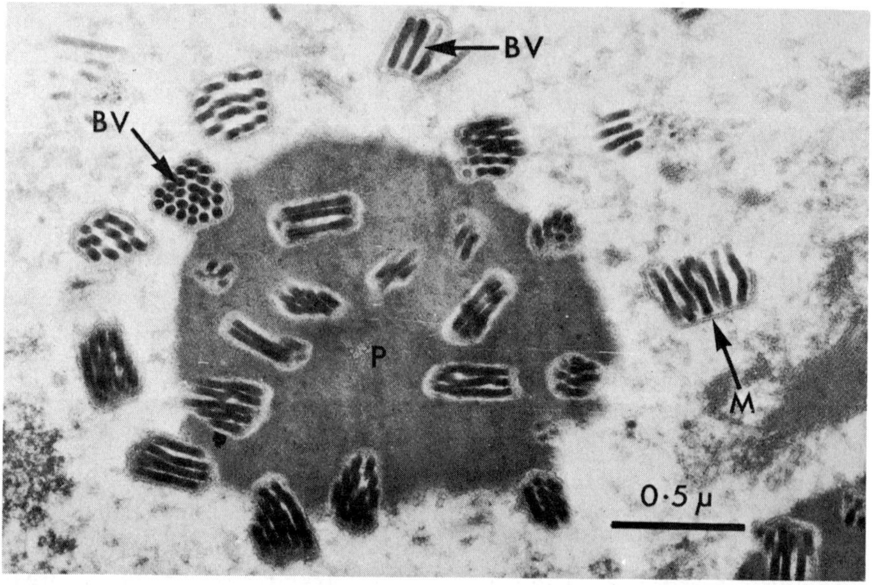

Fig. 3. Development of nuclear polyhedrosis virus in the eastern hemlock looper, *Lambdina fiscellaria fiscellaria.* Rod-shaped virions are produced in the nucleus of the infected cell; they aggregate in bundles (BV), are surrounded by a membrane (M), and eventually are embedded in protein that forms the polyhedra (P).

Insect viruses are usually very highly host specific. This gives them the same advantage enjoyed by the bacterial preparations, in that parasites and predators are unharmed, and there is no hazard for birds, fish, and other wildlife. True, only one pest insect will be controlled by one virus but, particularly in forest conditions, frequently only one species is troublesome at any given time. Should two or more require control, there would seem to be no reason why mixed viruses should not be applied.

Still another favourable characteristic of viruses, which has been demonstrated on numerous occasions, is their ability to create an epidemic and spread throughout a population. The first carefully studied case of this kind was the collapse of the European spruce sawfly outbreak in New Brunswick during the years 1937 to 1940. This was brought about by the accidental introduction of a virus, presumably as a result of including diseased cocoons in collections distributed for parasite introductions. The parasites played some part, but it seems clear that the virus was the crucial factor. In the 30 years since, the sawfly has never posed any problem in Canada, although it still occurs and has spread westward through Ontario. During the same period deliberate dissemination of virus in other sawfly populations has given similar results.

In agricultural crops too this phenomenon occurs. For example, as a result of migration of moths from more southerly areas, the cabbage looper each year is a major pest in parts of Ontario where pickling cabbage is grown, and each year toward the end of the season the population is practically wiped out by a virus disease. It has been shown that the inoculum for this disease persists in the soil from year to year (Jaques, 1970). But since it is a true infection, and since up to ten days may be required for an insect to acquire the infection and die so that it in turn provides a source of infection for others, extensive damage occurs before control is effected. Again field tests have shown that if the inoculum is artificially distributed on the plants early in the season, so that all or most of the first generation larvae from the migrants become infected, damage is greatly reduced (Jaques, personal communication).

Not all viruses are highly pathogenic. By coincidence those first carefully studied—nuclear polyhedrosis virus of the European spruce sawfly (Balch and Bird, 1944), of the European pine sawfly (Bird, 1953), and of the alfalfa caterpillar (Steinhaus, 1948)—are among the most virulent. Others are much less effective, such as those of the jackpine budworm and of the hemlock looper. The reasons for this difference in virulence are not clear. It is not related to the kind of virus, because extremes are found in the major types. Recently Bird (unpublished) has found a new form in the spruce budworm which he is calling "Y" virus until it has been characterized sufficiently to permit proper classification. He finds that it is highly infectious and highly lethal when sprayed on artificial diet on which the larvae are reared in the laboratory. This virus has an unusual effect in that it may prolong larval development so that infected individuals grow to a size much larger than normal. We are planning field tests with this virus during the coming summer.

One feature common to all insect pathogens so far studied is that the insects appear to be incapable of building up resistance to them, either by a selection process such as occurs with many chemical insecticides, or by the development of an immunity as in many diseases in higher animals. The European spruce sawfly population in New Brunswick, first infected by virus some 35 years ago—and there are two generations each year—still appears to be as susceptible as it was at first (Bird and Burk, 1961). The European pine sawfly has been under observation in Canada for 20 years and there is no evidence of resistance. And there are no reports of any decrease in susceptibility of any species to *Bacillus thuringiensis*.

As a final point in favour of pathogens, and one that is of increasing interest and concern at the present time, one must mention that all of them are of natural occurrence in the environment. Therefore they cannot be classed as pollutants in the sense that manufactured chemicals are. Pathogens are organic entities, and as such are subject to biodegradation and re-incorporation into the natural cycle. It must be admitted that there are very few specific studies of this point yet, but they will be done if only to meet the requirements for registration when pathogens are made available commercially. But observation and logic certainly point to this conclusion. Consider the biomass in a locust outbreak, for example, that is controlled by a fungus disease. Yet birds and other animals wander over the fields and feed on the remaining crops with no ill effects, and the next season everything is back to normal.

Regarding the potential for pathogens in the regulation of insect populations, there can be no conclusion other than that it is very great and awaiting development, but there are a number of obstacles that must be cleared away by further research. Fungi, although on occasion very effective in natural epizootics, are not yet manageable. There is insufficient knowledge about their physiology and development. Some of the recognized pathogenic forms, such as members of the genus *Beauveria*, are much less selective than would be desirable. Others, in the genera *Entomophthora* and *Massospora*, have restricted host ranges and appear to be good prospects for development.

Protozoa probably will not be particularly valuable in reducing outbreaks, but they undoubtedly play an important part in regulating population build-up. As techniques for producing and distributing them are developed it can be expected that they will be subject to some management and will be of considerable value.

Bacteria are already in commercial production and widely used. One, *Bacillus popilliae*, is specific for the Japanese beetle and although it is initially slow in action it tends to become established at an enzootic level and provides a continuing regulation of the population. The *Bacillus thuringiensis* preparations, on the other hand, are very rapid in their initial effects. Within hours, larvae cease feeding due to the action of the toxin. However, the total effect is more like that of a chemical, and no epizootic develops to exert a continuing regulation. At the present time these preparations appear to be most promising for agricultural use and especially on food crops where chemical residues must be kept very low or avoided completely, and where repeated applications are economically acceptable. They may also be useful in special forest situations such as in parks, fishing areas, water supply drainages, and so on. Compared to chemicals they are expensive, but as new and more virulent strains are discovered, permitting lower rates of application, this disadvantage will be overcome. It seems reasonable to expect that new species or varieties will be found that will be infectious rather than or in addition to being toxic, as has already been reported by Talalaev (1960) in Russia, and so will initiate epizootics thus making them of continuing value.

Taking all factors into consideration it appears that the viruses offer the greatest potential among the pathogens. Recognizing that they are not all equally virulent and that probably only the most virulent will be worth developing, their selectivity and their ability to spread through a population, and to persist and be transmitted to successive generations, are of very great importance. At present it almost appears that their high virulence may to some extent militate against them—the quantities required are in some cases so small that they do not offer attractive possibilities for commercial exploitation. The viruses seem to be particularly suited for control of forest pests where light populations can be tolerated, thus providing a reservoir of the pathogen that hopefully will increase if there is any tendency for a pest build-up.

Pathogens, like chemicals, are most effective when populations are high, but ways may be found to increase their impact. For instance, it may be possible to artificially infect an area in which a population build-up is expected so that an early epizootic will occur and an outbreak be prevented. This approch to spruce budworm control is currently being explored in Canada. It is reasoned that since outbreaks frequently seem to develop in and spread from a so-called epicentre, if this centre can be recognized and given early control treatment a widespread outbreak can be avoided. Theoretically an infectious pathogen introduced early would be ideal in such a situation.

Summary

In practice it is unlikely that very many pests will be controlled by use of pathogens alone. Rather we should take advantage of their selectivity to reduce the pest and at the same time leave parasites and predators, as well as other useful insects, unharmed, leading to an integrated control where poisonous chemicals are used only occasionally and sparingly.

Preparations such as *Bacillus thuringiensis*, whose effect is similar to that of a chemical in that it is quick but not self-perpetuating, will be of increasing value for agricultural crops because their use dose not result in a residue problem. Their cost will for the near future at least make them less acceptable for forest spraying except perhaps in special circumstances.

Viruses, which are rather slow in their initial effect but which frequently will spread throughout an area and persist through successive generations, will probably find their greatest use in the forests where limited populations can be tolerated, thus permitting the establishment of a balanced ecology at an acceptable level. If due to natural fluctuations the pest shows evidence of developing to unacceptably high population levels, artificial dissemination of additional virus would be preferable to intervening with non-selective chemicals.

Fungi are still unmanageable. Although on occasion they are sometimes extremely effective by natural development, much more needs to be known of their physiology and the factors that limit and control their growth and spread. This knowledge will be obtained, and when it is the often observed capabilities of these pathogens will be exploited effectively.

Microsporidia and other protozoa will probably not become agents to control outbreaks. Rather they will be exploited by establishing them in burgeoning populations so that they can exert their debilitating effects on the pest, especially by reducing the normal reproductive potential. In other words they will find their greatest use as preventive rather than cure.

In sum, it seems safe to say that the potential for pathogens in regulating insect populations is very great. Much progress has been made in the quarter century since insect pathology became established as a subject for intensive study and exploitation. Much more research is needed, but there appears to be every reason to expect that eventually pathogens will be agents of major importance, and in concert with other natural control factors they will greatly reduce our present almost total dependence on chemical pesticides for insect control.

Literature Cited

Balch, R. E. and F. T. Bird. 1944. A disease of the European pine sawfly, *Gilpinia hercyniae* (Htg.) and its place in natural control. Sci. Agr. 25: 65–80.

Bird, F. T. 1953. The use of a virus disease in the biological control of the European pine sawfly, *Neodiprion sertifer* (Geoffr.). Canadian Ent. 85: 437–446.

Bird, F. T. and J. M. Burk. 1961. Artificially disseminated virus as a factor controlling the European spruce sawfly, *Diprion hercyniae* (Htg.) in the absence of introduced parasites. Canadian Ent. 93: 228–238.

Jaques, R. P. 1970. Natural occurrence of viruses of the cabbage looper in field plots. Canadian Ent. 102: 36–41.

Steinhaus, Edward A. 1948. Polyhedrosis ("Wilt disease") of the alfalfa caterpillar. J. Econ. Ent. 41: 859–865.

Talalaev, E. V. 1960. The bacteriological method of control of *Dendrolimus sibericus*. Trans. Int. Conf. Insect Pathol. Biol. Control, Prague 1958, pp. 51–57.

Thomson, H. M. 1960. The possible control of a budworm infestation by a microsporidian disease. Canada Dept. Agr. Forest Biol. Div. Bi-Mo. Prog. Rept. 16 (4): 1.

Weiser, J. 1957. The possibilities of biological struggle with *Hyphantria cunea* Drury. III. Czechoslovakiz Parasitol. 4: 359–367.

Descriptors: insect pathogens; fungi; protozoa; bacteria; viruses; pest management with pathogens.

PLANT-INSECT INTERACTION

ANIMAL POPULATIONS, PLANT HOST RESISTANCE AND ENVIRONMENTAL CONTROL[1]

DAVID PIMENTEL

INTRODUCTION

FOLLOWING THE initial success of DDT after World War II, the widespread use of this and other organic pesticides became increasingly popular. Heavy public pressure was placed on the biologists responsible for pest control to devise better methods for the protection of man's crops, his health, and natural resources. Under these pressures economic biologists were forced to rely on empirical controls, using chemicals which killed the highest percentage of the pest population.

With the subsequent spectacular pest kills, public reliance upon pesticides increased. Even after the widely publicized warnings against pesticides in Carson's book *Silent Spring* (1962) and recommendations made by several Presidential Task Forces, public usage of pesticides for pest control more than doubled in the last 10 years! Today, more pests are controlled, more lives are saved, and more crops and natural resources are protected by pesticides than by all other means of pest control combined. Thus, this is a most important time to take stock of where we are today and where we are headed in general in pest control.

Examining pest control procedures, we find that balanced against the significant benefits of pesticides are the hazards of environmental pollution. Certainly pesticides are proven to be valuable tools in pest population control and will continue to be in the future. In addition, however, the many possibilities which exist in the development and use of environmental controls must be fully explored.

[1] This research was supported in part by National Science Foundation Grant (GB 7757, Environmental Biology).

TALL TIMBERS CONFERENCE ON ECOLOGICAL ANIMAL CONTROL BY HABITAT MANAGEMENT, 1969, pp. 19-28.

GENERAL CHARACTERISTICS OF NATURAL POPULATIONS

Understanding the basic characteristics of most natural populations is a starting point for any consideration of effective control methods. Some of the major points of interest are as follows:

A. *Animal populations in natural biotic communities are relatively stable.* Outbreaks in animal populations are generally rare in those natural biotic communities which have been left undisturbed by man. Constancy is characteristic of populations in natural communities. MacFadyen (1957) wrote that "it is generally agreed that the same species are usually found in the same habitats at the same seasons for many years in succession, and that they occur in numbers which are of the same order of magnitude." Without presenting the volumes of supporting data, it can be stated in summary that most changes in population numbers are mere ripples compared with the waves that are possible.

B. *Most species are rare in relative numbers.* If the numbers of a species are related to their food resource, the numbers of most species are classified as rare. To generalize further, in the Origin of Species, Darwin wrote, "rarity is the attribute of a vast number of species of all classes, in all countries."

C. *The majority of animals feed on living matter as opposed to non-living.* In nature, the majority of all animals may be classified as herbivore (grazer on living plants), parasite, or predator; few species are truly saprophytic. Although many animals are associated with dead matter, most of these animals are not true saprophytes but are herbivores feeding on bacteria, fungi, and other minute organisms present in the decaying matter.

D. *Population outbreaks frequently occur with newly introduced animal species.* The plant hosts on which the newly introduced animal is feeding often lack resistance to it. Under such conditions, the animal often increases rapidly on its new susceptible host and reaches outbreak levels. The introduction of the gypsy moth is one memorable example of this. The moth was brought to this country nearly a century ago by a French scientist who brought the eggs from France to his home in Medford, Massachusetts. Some moths escaped during a wind storm and within a short span of time these moths reproduced until an explosive outbreak took place in eastern Massachusetts. This moth destroyed many trees, especially the most susceptible oaks, and the gypsy moth still remains a relative serious pest in New England and surrounding states.

E. *Resistant factors which limit the feeding of animals on host plants are common in nature.* Various kinds of resistant factors occur naturally in plants and appear to be quite prevalent. The spines and thorns of many plants, such as cacti, gorse, and hawthorne, prevent or inhibit intense feeding by browsing animals. Toxins or growth inhibitors which are found in many kinds of plants limit animal feeding. These would include tannins in oak leaves, cyanide in bird's-foot trefoil, and nepatactone in catnip.

POPULATION REGULATORY MECHANISMS

Population numbers are limited by many factors in nature including competition and environmental discontinuity. In addition, I have proposed that the level of host resistance is an important factor in animal population control. This is a natural control mechanism I have termed the genetic feedback mechanism because of the way it works (Pimentel 1961). With this mechanism in play, population numbers (herbivore, parasite, or predator) are regulated in this way: High animal numbers exert strong selective pressures on their host-plant populations; the resulting selection alters the genetic make-up of the host population to make the host more resistant to attack; this in turn feeds back and the presence of more resistant-type hosts results in a lessening of the feed pressure of the herbivore. After many such cycles, the numbers of the animal herbivore population are limited to a point where relative stability results.

Through the functioning of the genetic feedback mechanism, resistant factors in the host plant can control the animal's feeding pressure and also the numbers of animals. The effectiveness of genetic change in a plant population in controlling animal numbers is well illustrated by the interaction of the resistant wheat crop in

Kansas and its Hessian fly pest. At first the Hessian fly population increased rapidly on susceptible wheat and soon became a serious pest of the wheat crop. Subsequently, with the late R. H. Painter's (1951) development of a resistant variety of wheat, reproduction of the Hessian fly on the resistant wheat dropped to an estimated level of less than one per parent fly and soon the fly population declined to a low level. Hence, by manipulation of genes naturally present in wheat, the Hessian fly population was reduced to a new low-level.

Genetic change in a plant population usually requires several generations to attain. As shown with the Hessian fly, the effects on the animal population can be dramatic. Let us examine how this genetic mechanism might function. When a large animal population exerts a severe feeding pressure on the plant population, large numbers of plants are destroyed. Obviously, the first plants destroyed are those most susceptible to the feeding pressure of the animal; the surviving plants carry one or more resistant genes. Under natural conditions the genetic change taking place in the plant population rarely would be due to mutations, but rather to recombination of the genes already present. That is, resistant genes already existing in the plant population at relatively low frequency would be recombined and concentrated in the succeeding generations of the plant population. This evolutionary procedure of recombination and concentration proceeds slowly. Because resistance is normally due to a combination of factors, it is usually polygenic in nature.

To illustrate, we could assume there might be 20 loci (2 genes per locus) for resistance in the host plant for some resistant character like hardness. As the susceptible genes at such locus are slowly replaced with resistant genes, the amount of resistance gradually increases until it is highly effective against the attacks of the pest.

The level of resistance in the host plant can be evaluated from the response of the animal pest to the change. The influence of increasing host resistance in the animal pest may be measured by decreased number of eggs produced, prolonged growth and development, decreased longevity and increased mortality. Ultimately the level of host resistance would result in the animal pest population having a lower birth than death rate and there would be a noticeable decline in the animal population until it becomes relatively sparse. At this stage, when animal numbers are rare, especially compared with plant biomass, the animal population would be removing only "interest" (excess energy or individuals) from the plant population. Then relative equilibrium would exist in the supply-demand relationship between plant and animal. The animal population would no longer be removing "capital" (that energy or those individuals, or both, needed for growth and maintenance and reproduction of the plant population). Evolution of this kind, with a balance between supply and demand, is possible when the genetic feedback mechanism functions.

A biological model of the functioning of the feedback based on systems analysis is typified in Figures 1, 2, and 3. When the animal is newly associated with its plant host the animal population soon reaches outbreak levels and fluctuations are relatively intense (Fig. 1). As resistant factors accumulate in the host plant (Fig. 2) the birth rate of the animal decreases. Associated with this change is a general decline in the density of the animal population and a dampening in the amplitude of the fluctuations (Fig. 1). During this stage a relative balance exists in the supply-demand relationship between plant and animal.

The decrease in amplitude of the fluctuations is due directly to a decline in reproductive rate of the animal (Fig. 3). An animal population which is increasing rapidly on a susceptible host will increase well beyond the carrying capacity of its host food supply. Populations over-shoot the carrying capacity of their hosts because there is a time lag in the control effects which limit animal numbers. For instance, it may take another generation or two for the control factor to become effective on the sensitive stage of the life history of the attacking animal. The large number of animals accumulated above the carrying capacity will severely deplete the food supply and result in a crash of the animal population (Fig. 3). When the animal is limited to "interest feeding" and the rate of increase is low there is no danger of large increases and large declines.

The rate of increase, as mentioned, is related to over-shooting the carrying capacity and the intensity of population fluctuations and this rate of increase is directly influenced by resistant factors in the

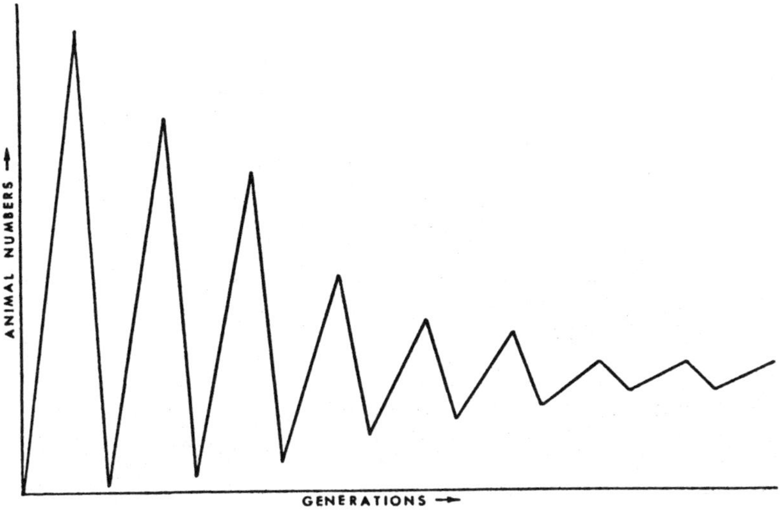

Fig. 1. Population trends in animal species which is newly associated with its plant host. Eventually as the two species evolve and become integrated with one another in the food chain of the community, the fluctuations decrease in intensity and the mean number of animals attacking the plant population will decrease. During the final stage of the association the animal population is feeding only on "interest" of the plant population.

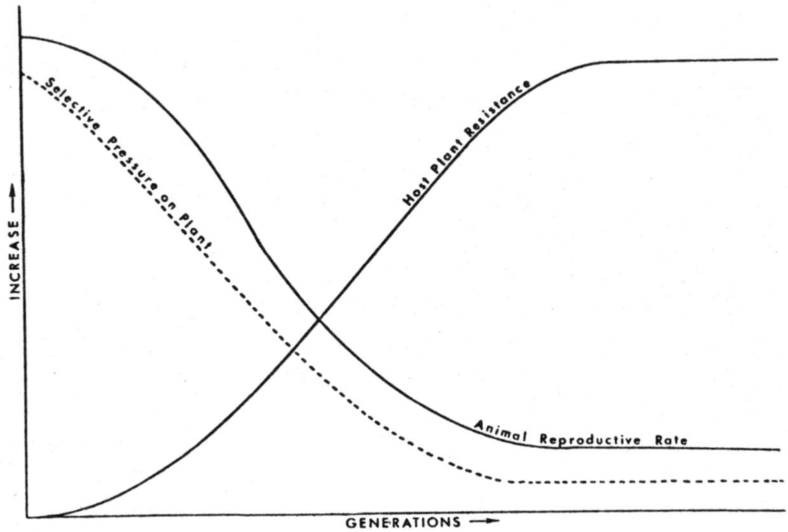

Fig. 2. Characteristic changes taking place in the newly associated animal and plant populations.

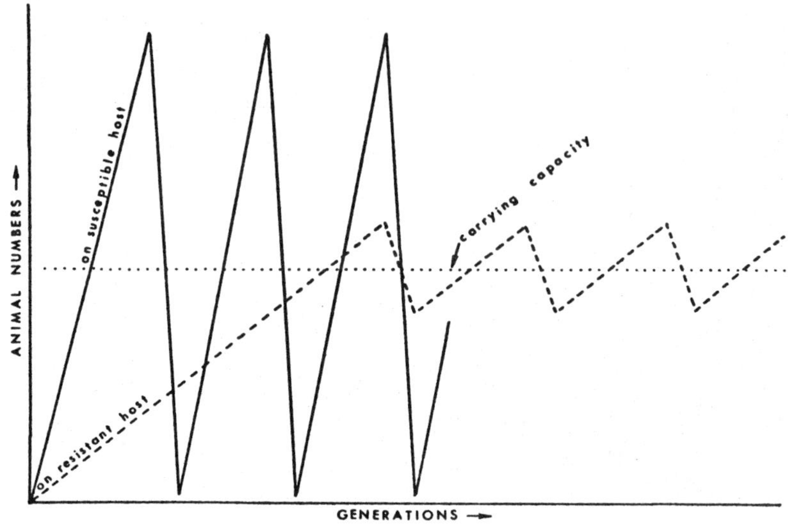

Fig. 3. Intensity of animal fluctuations related to rates of increase susceptible and resistant hosts.

plant (Fig. 2). For instance, when young pea aphids were placed on a common crop variety of alfalfa they produced a mean of 290 offspring in 10 days, whereas the same number of aphids feeding for a similar period on resistant alfalfa variety produced a mean of only 2 offspring (Dahms and Painter 1940). In another study, the mean rate of oviposition (eggs per generation) of the chinch bug on a susceptible strain of sorghum was 100, whereas on a resistant strain the mean oviposition was 0.9 (Dahms, 1948).

EVIDENCE SUPPORTING THE GENETIC FEEDBACK MECHANISM AS A CONTROLLING FACTOR IN NATURAL POPULATIONS

The data document that resistance is an effective factor in limiting animal numbers but how do resistance and genetic feedback relate to the evidence concerning natural populations? What about the outbreaks which frequently occur in newly introduced species? Meeting with little or no resistance, the foreign species increase rapidly on their susceptible plant host until resistance in the plant host gradually increases to an effective level. Both outbreaks and intense fluctuations will typify the newly introduced animal population. When relative stability is eventually reached and host resistance is effective in limiting animal numbers, the number of pest species will be low relative to the plant host. At that time the animal population is limited to feeding only on "interest" produced by its host plant. This is undoubtedly one of the important reasons why most animals are rare and especially rare relative to their food host. Animals which survive and multiply without destroying their host food supply obviously have greater chances for survival.

The responsiveness of living systems to environmental change and pressures and their ability to evolve stability is believed to account, in part, for the fact that most animals feed on living matter. The interaction between the eating and eaten species in the genetic feedback mechanism forms a fully responsive, evolving system. Living systems respond to change and can evolve into a balanced supply-demand system within the community.

The adaptation of supply by the plant host to the demand of the animal is fundamental to the stable community system. The plant host only responds and evolves resistance to the feeding pressure of the animal if the animals are too numerous and are taking more than "interest" from the plant host. A certain amount of selection is necessary to maintain the resistant characters in the plant host population but this selective pressure is relatively low. These principles all apply equally to parasite-host and predator-prey interactions and in fact determine the basic structure of communities. From this it is understandable why Elton's (1927) statement, "the whole structure and activities of the community are dependent upon the questions of food supply" takes on great significance in developing sound methods of population control.

PLANT RESISTANCE AS A MEANS OF PEST CONTROL

Just as plant resistance plays a significant role in animal population control in nature, the development of resistance in crop plants and livestock can be as effective for pest control. The few outstanding instances of control employing plant resistance have been the control of the Hessian fly, grape phylloxera, and spotted alfalfa aphid. However, compared with plant pathologists who have employed plant resistance for control of many crop diseases, entomologists have not fully explored the possibility of using plant resistance as a pest control measure. Plant resistance provides permanent long term control without any danger of pollution. To date there have been no failures in using plant resistance for insect control when it has been employed.

In the past, resistant genes occurring naturally in the host plant have been frequently lost as crop plants have been bred for yield. With emphasis on yield, insect resistance is usually ignored. The plants under selection for yield are protected by pesticides or screening to avoid insect damage, and this results in the loss of resistant genes in the breeding program. Therefore, whether we are breeding seedling trees or cabbages a real advance in pest control could be made if plant breeders would judge the performance of the new variety against the field environment (including insect pests) in which the plants will be grown commercially.

As mentioned, breeding for plant resistance does take considerable time, but the results more than pay the cost. In fact, plant resistance as a form of pest control offers one of the best cost/benefit ratios in the field of pest control techniques. The results thus far document a return of better than a $20 return for every dollar invested in control. Research for development, the largest single expense, is included in this cost figure. Contrasting pesticides, in which the largest expense is the application cost, the general return is estimated as about $4 for every dollar invested in control.

In addition to a most favorable cost-benefit ratio for control, one of the great advantages of plant resistance is its compatability with other means of control, and especially pesticides. Plant resistance is not affected by pesticide treatments as are parasites and predators. Further, the resistance in the plant does not have to be 100 percent for it to be used as an effective control. For example, if the resistance in a plant host were high enough to reduce the reproductive rate of the insect pest by 50 percent, then the usual pesticidal treatments could be decreased several fold. Not only would this be an advantage in cost, but it would also reduce pollution and lessen the chance for the development of pesticidal resistance which we all recognize is a serious problem in pest control today.

In conclusion, man should take a lesson from nature and employ host resistance more effectively for pest control. The full benefits of employing host plant resistance for pest control have yet to be realized.

LITERATURE CITED

Carson, R. 1962. Silent Spring. Houghton Mifflin, Boston. 368 pp.

Dahms, R. G. 1948. Effect of different varieties and ages of sorghum on the biology of the chinch bug. Jour. Agr. Res. 76: 271-288.

Dahms, R. G. and Painter, R. H. 1940. Rate of reproduction of the pea aphid on different alfalfa plants. Jour. Econ. Ent. 33: 482-485.

Elton, C. 1927. Animal ecology. Sidgwick and Jackson Ltd., London 209 pp.

MacFadyen, A. 1957. Animal Ecology. London, Pitman.

Painter, R. H. 1951. Insect resistance in crop plants. New York, MacMillan.

Pimentel, D. 1961. Animal population regulation by the genetic feedback mechanism. Am. Nat. 95:65-79.

ECONOMIC THRESHOLD

Defining the Economic Threshold

J. C. Headley

Economic entomology as a discipline is concerned with the application of entomological knowledge for the purpose of providing economic benefit. This application may take the form of either preventing insect damage to man and his property or by producing beneficial entomological effects such as pollination and honey production.

Historically, economic entomologists have concerned themselves with preventing damages from insects. From this concern has come the concept of the economic threshold, sometimes referred to as the economic injury level. While the concept has received considerable attention, its definition has not been rigorous and has, therefore, been ambiguous. The purpose of this paper is to develop a definition of the economic threshold that will hopefully be a rigorous one, with a minimum of ambiguities.

SOME PREVAILING DEFINITIONS

Reference to economic thresholds is found frequently in the entomology literature. Edwards and Heath (1964) say that a pest population has reached the economic threshold when the population is large enough to cause damages valued at the cost of practical control. Bierne (1966) suggests that the economic threshold is a critical level of damage above which damages will not be tolerated. He continues to say that damages above this critical level cause the pest responsible

PEST CONTROL STRATEGIES FOR THE FUTURE NATIONAL ACADEMY OF SCIENCES, 1972, pp. 100-108.

to be labeled as a damaging pest and consideration is given or action is taken to try to abate the damage. The Subcommittee on Insect Pests of the Committee on Plant and Animal Pests established by the National Research Council (National Academy of Sciences, 1969) defined the economic threshold as "the level at which damage can no longer be tolerated and, therefore, the level at or before which it is desirable to initiate deliberate control activities." They go on to state that "the definition should probably be amended to consider a more critical density as that where the loss caused by the pest equals in value the cost of available control measures." Their definition is further tempered by including for consideration, adverse side effects which may entail costly ecological consequences. Obviously, the threshold should also include beneficial side effects that might be involved with predators or parasites of beneficial species.

The definitions provided above do not include any statement of the unit to which the definition applies, that is, a field as contrasted to a farm, or an individual producer as contrasted with a community or region. According to the statements cited, the implication is that the economic threshold is one of the components of a decision model for individual producers considering pests one at a time at a given point in time. To the extent that individual producer's actions are not independent of effects felt by other producers or are not independent as between time periods, the aggregate result of decisions based on this concept may not be consistent with expected micro-results.

ANOTHER DEFINITION

The concept of an economic threshold in its simplest form seems to involve about three variables. These are damage, pest population, and, perhaps, time. These variables are related in that damage is a function of pest population and pest population is a function of time.

As the first suggestion, conceive of a damage function at time t causally dependent upon the pest population in time t, such as:

$$D_t = bP_t^2 - A. \qquad (1)$$

where D_t = damage in t
P_t = pest population in t
A = constant to define the damage tolerance level

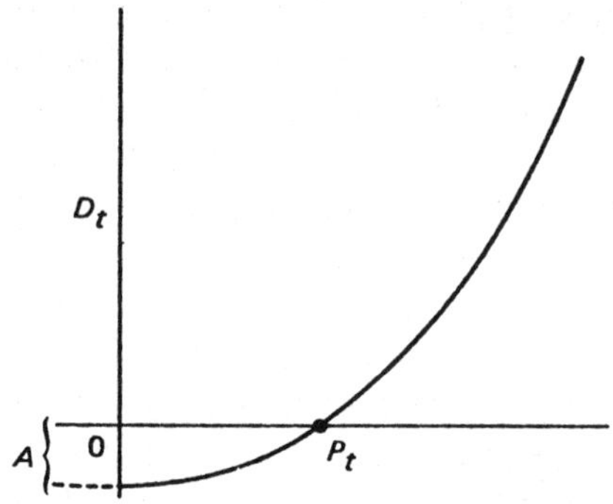

FIGURE 1 Product damage related to pest population.

b = constant parameter which enters into the incremental damage resulting from P_t.

The graph of this function is shown in Figure 1.

Assume that the population, P_t, is related to previous population levels at some point, say, P_{t-n}. That is, at a point n time periods before t, which is entomologically significant, there is a population that will grow to P_t. These populations are related by the function:

$$P_t = P_{t-n}\,(1+r)^n \tag{2}$$

where P_{t-n} = the population n periods prior to t
r = growth rate of the population per time period
$(1+r)^n$ = compound growth factor.

The value for P_t in (2) can be substituted into (1) to derive D_t as a function of P_{t-n}.

$$D_t = b\left[P_{t-n}\,(1+r)^n\right]^2 - A \tag{3}$$

What is present here, then, is a relationship where the population at time period $t–n$ is the present value of the population at time period t. Therefore, if, through control activities, the population at $t–n$ is maintained at some level, the population at t is determined and, therefore, the damage in t is determined.

Now, if the product yield is some deterministic biological function of D_t such as

$$y = N - cD_t \tag{4}$$

where y = product yield
c = a constant parameter measuring incremental yield effects
N = a constant measuring yield with no damage, and D_t is defined as before

then, the effects of population on yield can be evaluated. Because by substitution

$$y = N - c\left\{b\left[P_{t-n}\,(1+r)^n\right]^2 - A\right\}. \tag{5}$$

For an individual farm producer who is a price taker, then the yield function (5) can be multiplied by a constant price per unit R to provide value of production as a function of the population at the point $t–n$. This is shown in Figure 2. Note that the value function has

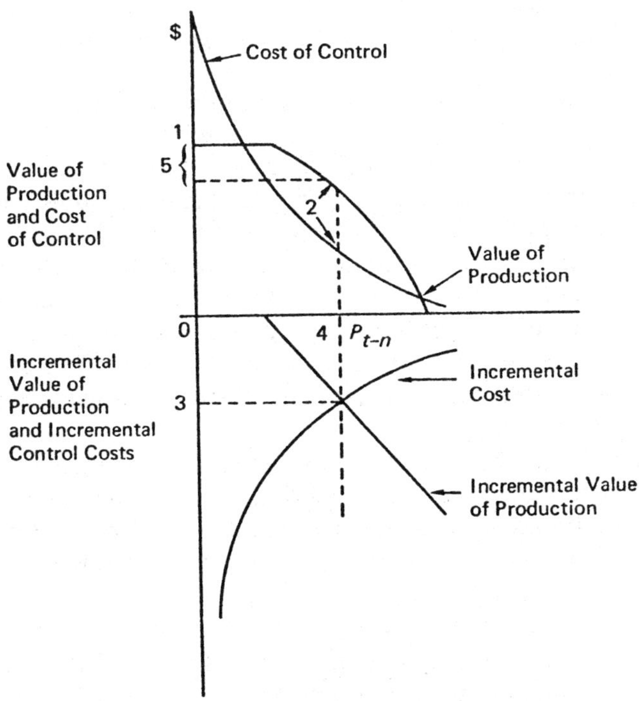

FIGURE 2 Relation of value of production, control cost, and pest population.

the same shape as (5) because of our assumption of constant price per unit. Figure 2 shows that as the population increases, the value of production is decreasing at an increasing rate that is consistent with our assumed damage function (1). When P_{t-n} reaches a value where the expression $cb\, P_{t-n}^2\, (1+r)^{2n} = N + CA$, then y = 0 and the value of production equals zero.

The final element in this model is a function that relates control costs to population, i.e., which tells the cost of control to maintain a population of certain size. Assume that this cost relation is continuous and decreases as the population is allowed to be larger and that this cost decreases at a decreasing rate, which is in keeping with the findings of entomologists that the cost of reducing a population increases at an increasing rate (National Academy of Sciences, 1969).

$$0 = \frac{L}{P_{t-n}} \tag{6}$$

where 0 = the total control cost
L = a constant parameter influencing incremental costs

and P_{t-n} is defined as before. This function is also graphed in Figure 2.

Now is the time to pose the crucial question. What is the economic threshold? A number of points are labeled in Figure 2. Point 1 is the normal producer revenue when the pest population is absent or statistically equal to zero. Point 2 identifies the points where the rate of change of the value of production function is equal to the rate of change of the cost of control function. The value of this rate of change is shown at point 3 on the incremental scale. Point 4 is the pest population level associated with point 2. Point 4, then, is the economic threshold!

This definition can be defended with the following argument: Because of the nature of the damage function, the incremental losses in value of production due to increases in population at low levels are small, while the incremental costs of reducing populations at these levels are high. This is shown by the relative slopes of the two curves in the upper quadrant to the left of point 4. As population increases, the costs of maintaining the population diminishes and the damage mounts, i.e., the value of production decreases. At point 2 on the curves or point 4 on the P_{t-n} axis the cost of reducing the population by one increment is just equal to the incremental increase in the value of production that results. Therefore, it is not advantageous to reduce the population below this level, but rather to sacrifice the value of production shown by point 5, since the costs of obtaining it, in this model, are astronomical.

Therefore, the economic threshold here defined becomes the population that produces incremental damage equal to the cost of preventing that damage. It is not a zero level of damage. Furthermore, the economic threshold population, here defined, also becomes the *optimal* population in an economic sense!

SOME CHARACTERISTICS OF THE DEFINITION

As defined above, the economic threshold has certain properties. First, the critical (optimal) population level will become smaller as the value of the product increases given constant control technology and constant prices for control inputs. This is shown in Figure 3. As the price per unit of the product increases, the slope of the value of production function increases. This merely means that resources can be

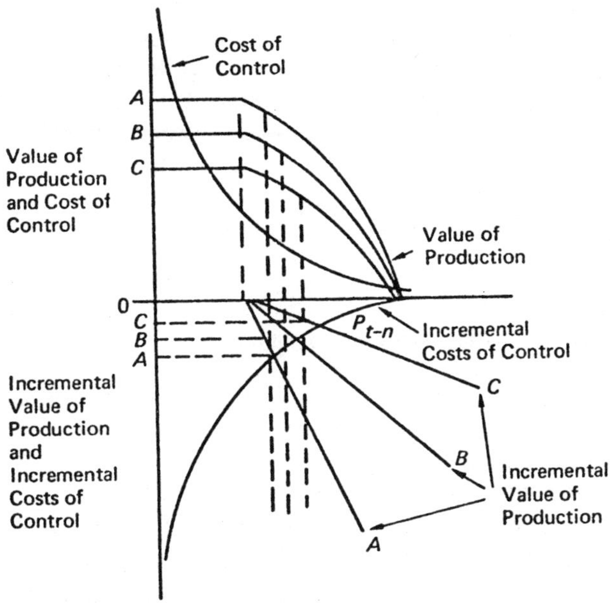

FIGURE 3 Relation of value of production, control costs, and pest population: product prices variable.

justified to prevent another increment of damage. Therefore, the economic threshold is a function of the product price. Similarly, increases in the unit cost of control methods will shift the control cost curve to the right and also increase the slope of the cost curve, which tends to move the economic threshold toward higher population levels.

Second, the definition given above applies only to an individual producer where the control measures are specific to a particular pest, a particular product, and a particular producer in a particular season. That is, any side effects at a point in time or through time are assumed to be absent. Such side effects could be included in the analysis by altering the control cost function.

Third, this model has assumed that the cost of control function is continuous, i.e., that the incremental costs can be defined at all points on the total cost function. This is not crucial to the findings, however, as long as the general level of control decreases as larger populations are allowed to survive. An example of the discontinuous cost of control function is shown in Figure 4. Here the cost of control function is a step function with a constant rate of change given by the incremental cost function. The increase in costs from level 1 to level 2 brings forth added production value measured by 3. The average rate of change in the value of production over control levels 1 and 2 occurs in the population range that occurs with control level 2 and, therefore, the added production exceeds the added costs. This is evident by the graph showing incremental values. Therefore, the same criterion applies as for the continuous model. The same conclusions regarding product price-changes and control inputs price-changes as in the continuous control cost case also apply.

SOME CONCLUSIONS AND IMPLICATIONS

Based on the definition of the economic threshold given in this paper the following conclusions can be made:

1. The economic threshold is responsive to and determined by prices of the product protected and the prices of control inputs.
2. It is entirely possible that the population level justifying control activities is larger than the population level where yield damage first occurs.

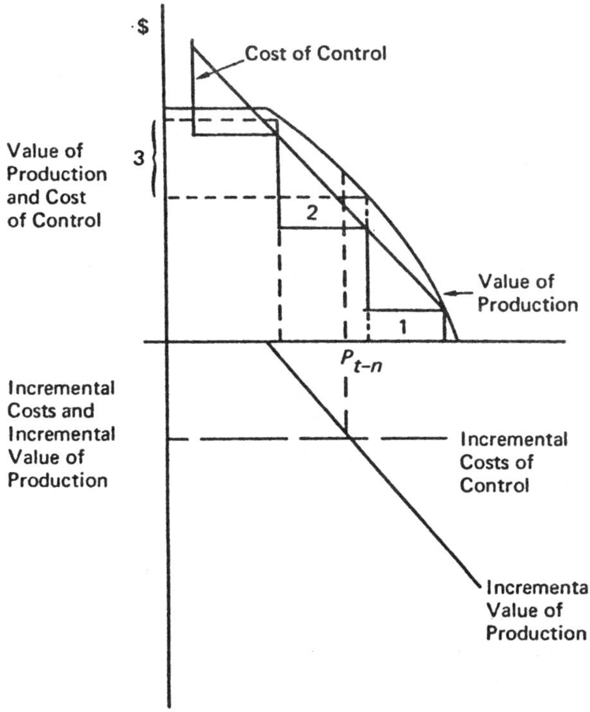

FIGURE 4 Relation of value of production, control costs, and pest population: cost of control discontinuous.

3. The economic threshold, when given this particular *economic* definition also becomes the optimal (net return-maximizing) level of pest population. No smaller or larger population will provide greater net returns above control costs.

4. Unless the incremental costs are less than the incremental value of damages prevented for the entire range of population levels, there is no economic justification for eradication policies under the assumption of time-independence built into this definition. The exception to this would be a guaranteed, once-and-for-all eradication, with a cost of maintaining zero level populations less than the costs of maintaining populations at greater than zero levels.

The implications of this definition of the economic threshold concern the need for research on the topic as a basis for establishing rational control programs and policies. There is a need for the development of damage functions for major pests of important products. Furthermore, control cost functions need to be developed for the same pests for alternative methods of control and combinations thereof.

Following the development of the information mentioned above, damage and cost functions over time are needed to evaluate the longer-term effects of control programs and to design programs that approach optimality over time.

Finally, further information is needed to determine the effects of control programs on overall populations in an area, target and non-target alike. Such research would evaluate strategies for maximizing community returns from pest control and would include information on the expected distribution of damages and benefits as a basis for broadening the support for control strategies and allocating the benefits and costs resulting therefrom.

REFERENCES

Beirne, Bryan P. 1966. Pest management. CRC Press, Cleveland.

Edwards, C. A., and G. W. Heath. 1964. The principles of agricultural entomology. Charles C Thomas, Springfield, Illinois.

National Academy of Sciences. 1969. Insect-pest management and control. *In* Principles of plant and animal pest control. Publ. 1695, Washington, D.C.

INTEGRATED CONTROL

THE INTEGRATED CONTROL CONCEPT

VERNON M. STERN, RAY F. SMITH, ROBERT van den BOSCH, and KENNETH S. HAGEN

All organisms are subjected to the physical and biotic pressures of the environments in which they live, and these factors, together with the genetic make-up of the species, determine their abundance and existence in any given area. Without natural control, a species which reproduces more than the parent stock could increase to infinite numbers. Man is subjected to environmental pressures just as other forms of life are, and he competes with other organisms for food and space.

Utilizing the traits that sharply differentiate him from other species, man has developed a technology permitting him to modify environments to meet his needs. Over the past several centuries, the competition has been almost completely in favor of man, as is attested by decimation of vast vertebrate populations, as well as populations of other forms of life (Thomas, 1956). But while eliminating many species, as he changed the environment of various regions to fit his needs for food and space, a number of species, particularly among the Arthropoda, became his direct competitors. Thus, when he subsisted as a huntsman or foraged for food from uncultivated sources, early man was largely content to share his subsistence and habitat with the lower organisms. Today, by contrast, as his population continues to increase (Hertzler, 1956) and his civilization to advance, he numbers his arthropod enemies in the thousands of species (Sabrosky, 1952).

The increase to pest status of a particular species may be the result of a single factor or a combination of factors. In the last century, the most significant factors have been the following.

First, by changing or manipulating the environment, man has created conditions that permit certain species to increase their population densities

HILGARDIA, 1959, vol. 29, pp. 81-101.

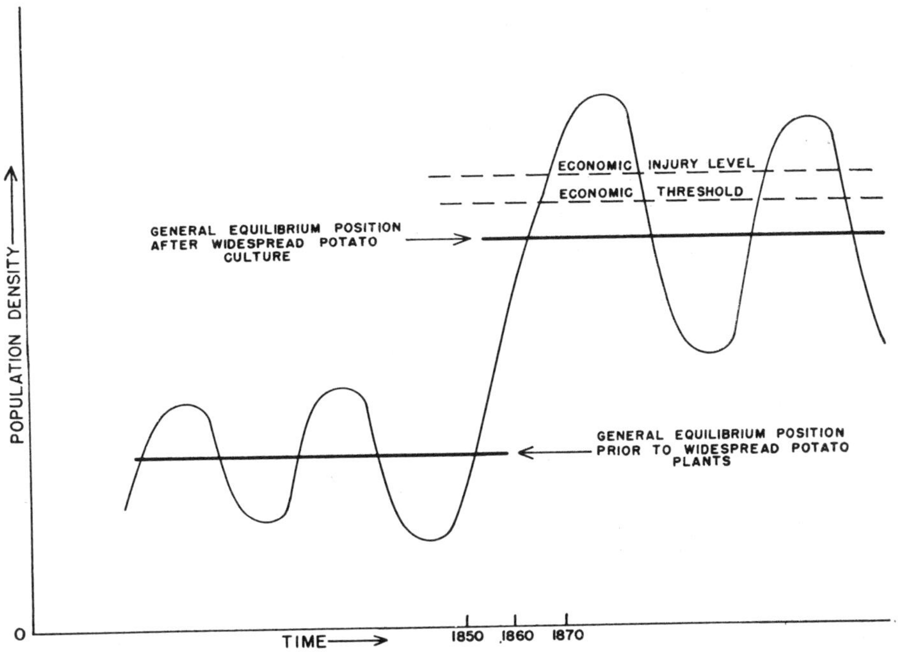

Fig. 1. Schematic graph of the change in general equilibrium position of the Colorado potato beetle, *Leptinotarsa decemlineata* following the development of widespread potato culture in the United States. For a discussion of the significance of economic-injury levels and economic thresholds in relation to the general equilibrium position, see p. 89.

(Ullyett, 1951). The rise of the Colorado potato beetle, *Leptinotarsa decemlineata* (Say), to pest status occurred in this manner (see fig. 1). When the potato, as well as other solanaceous plants, was brought under widespread cultivation in the United States, a change favorable to the beetle occurred in the environment, which enabled it to become very quickly an important pest (Trouvelot, 1936). Similarly, when alfalfa, *Medicago sativa* L., was introduced into California about 1850, the alfalfa butterfly, *Colias philodice eurytheme* Boisduval, which had previously occurred in low numbers on native legumes, found a widespread and favorable new host plant in its environment, and it subsequently became an economic pest (Smith and Allen, 1954).

A second way in which arthropods have risen to pest status has been through their transportation across geographical barriers while leaving their specific predators, parasites, and diseases behind (Smith, 1959). The increase in importance through such transportation is illustrated by the cottony cushion scale, *Icerya purchasi* Maskell (see fig. 2). This scale insect was introduced into California from Australia on acacia in 1868. Within the following two decades, it increased in abundance to the point where it threatened economic disaster to the entire citrus industry in California. Fortunately, the timely importation and establishment of two of its natural ene-

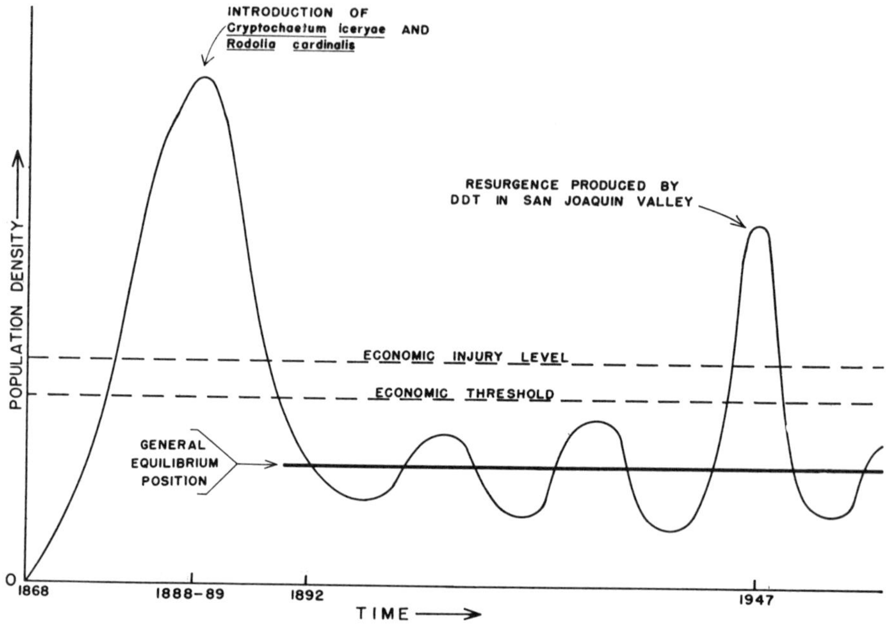

Fig. 2. Schematic graph of the fluctuations in population density of the cottony cushion scale, *Icerya purchasi*, on citrus from the time of its introduction into California in 1868. Following the successful introduction of two of its natural enemies in 1888 this scale was reduced to noneconomic status except for a local resurgence produced by DDT treatments.

mies, *Rodolia cardinalis* (Mulsant) and *Cryptochaetum iceryae* (Williston), resulted in the complete suppression of *I. purchasi* as a citrus pest (Doutt, 1958).

The cottony cushion scale again achieved the status of a major pest when the widespread use of DDT on citrus in the San Joaquin Valley eliminated the vedalia (Ewart and DeBach, 1947).

A third cause for the increasing number of pest arthropods has been the establishment of progressively lower economic thresholds (see p. 89 for definition and discussion). This can be illustrated by lygus bugs (*Lygus* spp.) on lima beans. Not too many years ago the blotches caused by lygus bugs feeding on an occasional lima bean were of little concern, and lygus bugs were considered a minor pest on this crop. However, with the emphasis on product appearance in the frozen-food industry, a demand was created for a near-perfect bean. For this reason, economic-injury thresholds were established and lygus bugs are now considered serious pests of lima beans.

In the face of this increased number of arthropod pests man has made remarkable advances in their control, and economic entomology has become a complex technical field. Of major importance have been new developments in pesticide chemistry and application.

The discovery of the insecticidal properties of DDT, and its spectacularly successful application to arthropod-borne disease and agricultural pest problems, spurred research in chlorinated hydrocarbon chemistry and stimulated the development of other organic pesticides. On a national scale, the experiment stations, state and governmental agencies, and commercial companies, all searching for new or better answers to old insect-pest problems, eagerly accepted the new chemicals. Within a short period many became an integral part of public health and agricultural pest-control programs. Without question, the rapid and widespread adoption of organic insecticides brought incalculable benefits to mankind, but it has now become apparent that this was not an unmixed blessing. Through the widespread and sometimes indiscriminate use of pesticides, the components and intricate relations of crop environments have been drastically altered, and as a result a number of serious problems have arisen (Wigglesworth, 1945; Michelbacher, 1954; Pickett, 1949; Pickett and Patterson, 1953; Solomon, 1953; and others). Among these new problems and old ones which have been aggravated are:

1. Arthropod resistance to insecticides. This phenomenon relating to the genetic plasticity of the arthropods has been reviewed by Metcalf (1955), Hoskins and Gordon (1956), Crow (1957), Brown (1959), and others. In many cases, resistance is already drastic enough to have eliminated certain insecticides from important pest-control programs. There are today in excess of 70 demonstrated cases of arthropod resistance. Actually, a much larger number of pest species exist which are developing resistance or have already done so, but there has not been time to evaluate these cases.

2. Secondary outbreaks of arthropods other than those against which control was originally directed (Massee, 1954; DeBach and Bartlett, 1951; Ripper, 1956; and others). These outbreaks usually result from the interference of the insecticide with biological control (Lord, 1947; Bartlett and Ortega, 1952; Michelbacher, 1954; Michelbacher and Hitchcock, 1958; and others). This may also occur through the effect of the insecticide on the plant, which, in turn, affects the development of the secondary pest (Fleschner and Scriven, 1957). An example is the increase in mites on plants growing in soil receiving certain chemical treatments (Klostemeyer and Rasmussen, 1953).

3. The rapid resurgence of treated species necessitating repetitious insecticide applications (Holloway and Young, 1943; Bovey, 1955; Schneider, 1955; Stern and van den Bosch, 1959; and others). These flarebacks occur from individuals surviving treatment or from individuals migrating into the treated area, where they can reproduce unhindered because their natural enemies have been eliminated.

4. The toxic insecticide residues on food and forage crops (Brown, 1951; Linsley, 1956). This problem may result from two sources. First, untimely applications or accidental increases in dosage may result in residues above the tolerance limits. Second, the first three problems mentioned above are interrelated and by aggravating one another may lead to excessive treatment and a residue problem. For example, where the level of resistance is increasing, it requires either more frequent applications or higher insecticide dosages to control the pest, or both. This increased insecticide program may in turn have a drastic effect on the ecosystem, which frequently results in outbreaks of secondary pests or rapid resurgence of the resistant pest

for which control was originally intended. Often, under such conditions, where insects threaten the crop or marketability of a crop close to harvest, the grower is faced with the problem of suffering a severe monetary loss or of making an insecticide application closer to harvest than is ordinarily permissible. In many instances, the end result is a residue far above the accepted tolerance limit at harvest time.

5. Hazards to insecticide handlers and to persons, livestock, and wildlife subjected to contamination by drift (Hayes, 1954; Petty, 1957; Upholt, 1955).

6. Legal complications from suits and other actions pertaining to the above problem.

Unquestionably, some of these problems have arisen from our limited knowledge of biological science; others are the result of a narrow approach to insect control. Few studies have included basic investigations on the effects the chemicals might have on other components of the ecosystems to which the pests belong. The entomologist may recognize the desirability of a thorough investigation of these aspects, but because of the need for immediate answers to pressing problems and because of other pressures, he does not have the necessary time. In other instances because fundamental knowledge is lacking, the investigator may be unaware of the intricate nature of the biotic complex with which he is dealing, and of the destructive potential that many chemicals in use today have on the environment of the pests. Finally, and most unfortunately, there are workers who are highly skeptical that biotic factors are of any consequence in the control of pest population densities and thus choose to ignore any approach to pest control other than the use of chemicals.

Whatever the reasons for our increased pest problems, it is becoming more and more evident that an integrated approach, utilizing both biological and chemical control, must be developed in many of our pest problems if we are to rectify the mistakes of the past and avoid similar ones in the future (DeBach, 1951, 1958*a*; Pickett, Putman, and Roux, 1958; Ripper, 1944; Huffaker and Kennett, 1956; Wille, 1951; Michelbacher and Middlekauff, 1950; and others).

TERMINOLOGY

To clarify the discussion in other parts of this paper some definitions and explanations of terms are here given:

Biological control. *The action of parasites, predators, or pathogens on a host or prey population which produces a lower general equilibrium position than would prevail in the absence of these agents.* Biological control is a part of natural control (*q.v.*) and in many cases it may be the key mechanism governing the population levels within the framework set by the environment. If the host or prey population is a pest species, biological control may or may not result in economic control. Biological control may apply to any species whether it is a pest or not, and regardless of whether or not man deliberately introduces, manipulates, or modifies the biological-control agents.

Biotic insecticide. *A biotic mortality agent applied to suppress a local insect pest population temporarily.* The effects of the agent usually do not persist and they are similar to those resulting from the use of a chemical insecticide in that they do not produce a permanent change in the general equilibrium position. A polyhedrosis virus applied as a spray to control the alfalfa caterpillar is a typical example of a biotic insecticide. Preparations of microörganisms used in this manner are sometimes referred to as *microbial insecticides.* Predators, such as lady beetles, or parasites, when they are released in large numbers, can also act, in some instances, as biotic insecticides.

Biotic reduction. *Deaths or other losses to the population* (e.g., *dispersal, reduced fecundity*) *caused or induced by biotic elements of the environment in a given period of time.*

Economic control. *The reduction or maintenance of a pest density below the economic-injury level* (q.v.).

Economic-injury level. *The lowest population density that will cause economic damage.* Economic damage is the amount of injury which will justify the cost of artificial control measures; consequently, the economic-injury level may vary from area to area, season to season, or with man's changing scale of economic values.

Economic threshold. *The density at which control measures should be determined to prevent an increasing pest population from reaching the economic-injury level.* The economic threshold is lower than the economic-injury level to permit sufficient time for the initiation of control measures and for these measures to take effect before the population reaches the economic-injury level.

Ecosystem. *The interacting system comprised of all the living organisms of an area and their nonliving environment.* The size of area must be extensive enough to permit the paths and rates of exchange of matter and energy which are characteristic of any ecosystem.

General equilibrium position. *The average density of a population over a period of time (usually lengthy) in the absence of permanent environmental change.* The size of the area involved and the length of the period of time will vary with the species under consideration. Temporary artificial modifications of the environment may produce a temporary alteration of the general equilibrium position (*i.e.,* a temporary equilibrium).

Governing mechanism. *The actions of environmental factors, collectively or singly, which so intensify as the population density increases and relax as this density falls that population increase beyond a characteristic high level is prevented and decrease to extinction is made unlikely.* The governing mechanisms operate within the framework or potential set by the other environmental elements.

Integrated control. *Applied pest control which combines and integrates biological and chemical control.* Chemical control is used as necessary and in a manner which is least disruptive to biological control. Integrated control may make use of naturally occurring biological control as well as biological control effected by manipulated or introduced biotic agents.

Microbial control. *Biological control that is effected by microörganisms (including viruses).*

Natural control. *The maintenance of a more or less fluctuating population density within certain definable upper and lower limits over a period of time by the combined actions of abiotic and biotic elements of the environment.* Natural control involves all aspects of the environment, not just those immediate or direct factors producing premature mortality, retarded development, or reduced fecundity; but remote or indirect factors as well. For most situations, governing mechanisms (*q.v.*) are present and determine the population levels within the framework or potential set by the other environmental elements. In the case of a pest population, natural control may or may not be sufficient to provide economic control.

Natural reduction. *Deaths or other losses to the population caused by naturally existing abiotic and biotic elements of the environment in a given period of time.*

Population. *A group of individuals of the same species that occupies a given area.* A population must have at least a minimum size and occupy an area containing all its ecological requisites to display fully such characteristics as growth, dispersion, fluctuation, turnover, dispersal, genetic variability, and continuity in time. The minimum population and the requisites in occupied area will vary from species to species.

Population dispersion. *The pattern of spacing shown by members of a population within its occupied habitat and the total area over which the given population may be spread.*

Selective insecticide. *An insecticide which while killing the pest individuals spares much or most of the other fauna, including beneficial species, either through differential toxic action or through the manner in which the insecticide is utilized (formulation, dosage, timing, etc.).*

Supervised insect control. *Control of insects and related organisms supervised by qualified entomologists and based on conclusions reached from periodically measured population densities of pests and beneficial species.* Ideally, supervised control is based on a sound knowledge of the ecology of the organisms involved and projected future population trends of pests and natural enemies.

Temporary equilibrium position. *The average density of a population over a large area temporarily modified by a procedure such as continued use of insecticides.* The modified average density of the population will revert to the previous or normal density level when the modifying agent is removed or expended (*cf.* "general equilibrium position").

THE NATURE AND WORKING PRINCIPLES OF BIOLOGICAL AND CHEMICAL CONTROL

Biological Control. Biological controls are part of natural control which governs the population density of pest species. On the other hand, with certain exceptions, chemical controls involve only immediate and temporary decimation of localized populations and do not contribute to permanent density regulation. This distinction is not always clearly made, and biological control is often thought of as being similar in its action to chemical control. Perhaps one reason for the misunderstanding is that in spectacular instances a biotic agent may act in the manner of a chemical in eliminating a local pest population. For example, this may occur when weather conditions are favorable and disease pathogens eliminate a localized pest population. Parasites and predators may sometimes act in a similar manner. However, the important prevailing characteristic of biological control is one of permanent population-density regulation. Usually these governing mechanisms occur over such a large area and are so subtle or intricate in their action that they are not easily observed and recorded; thus they tend to be overlooked.

A principal phase of applied biological control is the importation and establishment of natural enemies of pests that accidentally gain entry into new geographical regions. These new pests frequently escape the natural enemies that help to regulate their densities in the areas to which they are indigenous (Elton, 1958). Under satisfactory conditions in the new environment, the pest may flourish and reach damaging abundance. As a counter measure, the natural enemies are obtained from the native home of the pest and transplanted into the new environment to increase the biotic resistance of the environment to the pest.

Biological control is thus utilized to permanently increase environmental resistance to an introduced pest. The hope is that the introduced enemies will lower the general equilibrium position of the pest sufficiently to maintain it permanently below the economic threshold. Most often the introduction of a biotic agent is not so spectacular, and it is an exception when the general equilibrium position of the introduced pest is lowered sufficiently to prevent its occasionally or even commonly reaching economic abundance at certain times or places (Clausen, 1956; Simmonds, 1956). This, of course, is precisely the status of a native pest which, though attacked by a complex of parasites and predators, still has a general equilibrium position high enough to permit it occasionally to cause damage of greater or lesser severity. Thus, in any geographic area the governing mechanisms in the environment are constantly at work to counteract the inherent natality of plant and animal pest species. In terms of crop protection, these regulating factors actually keep thousands of potentially harmful arthropod species permanently below economic thresholds. Moreover, these environmental pressures tend to localize the outbreaks of those forms which on occasion are capable of rising above economic thresholds. A biological control agent is self-perpetuating and capable of response to fluctuations in the population density of the pest it attacks. Biological controls, whether imported or native, are permanent characteristics of a given environment.

Chemical Control. Chemical control of an arthropod pest is employed to reduce populations of pest species which rise to dangerous levels when the environmental pressures are inadequate. When chemicals are used, the damage from the pest species must be sufficiently great to cover not only the cost of the insecticidal treatment but also the possible deleterious effects, such as the harmful influence of the chemical on the ecosystem. On some occasions, the pest outbreak may cover a wide area; in other instances, damaging numbers occur in very restricted locations. These outbreaks occur during the season favorable to the pest, with the relaxed environmental pressures occurring some time before the outbreak. Chemical control is only needed at those times and places where natural control is inadequate. Chemical control should act as a complement to the biological control.

An insecticide must always be manipulated by man, who adds it to a restricted segment of the pest's environment to decimate a localized pest population. Because chemical insecticides are nonreproductive, have no searching capacity, and are nonpersistent, they constitute short-term, restricted pressures. They cannot permanently change the general equilibrium position of the pest population nor can they restrain an increase in abundance of the pest without repeated applications. Therefore, they must be added to the environment at varying intervals of time.

In certain pest-control programs, the insecticide is applied over extensive geographical areas. In some areas, after application, the pest population density may be far below the economic threshold and below its general equilibrium position; but since the insecticide is not a permanent part of the environment, the pest may return to a high level when the effects of the insecticide are gone.

The effectiveness of a chemical insecticide or a biotic insecticide is measured in per cent of kill or in per cent of clean fruits, uninjured cotton bolls, and so forth, in the area of application. Such applications have little influence on the pest in adjoining areas except as localized population depressants. In general, this contrasts sharply with the role of the permanent biotic mortality agent, whose effectiveness is best measured by its influence on the general equilibrium position of the pest species over an entire geographical region or a long period of time.

ECONOMIC THRESHOLDS AND THE GENERAL EQUILIBRIUM POSITION

Chemical control should be used only when the economic threshold is reached and when the natural mortality factors present in the environment are not capable of preventing the pest population from reaching the economic-injury level. The economic-injury level is a slightly greater density than the economic threshold. This difference in densities provides a margin of safety for the time that elapses between the detection of the threatening infestation and the actual application of an insecticide. The economic threshold and the economic-injury level of a pest species can vary depending upon the crop, season, area, and desire of man; the general equilibrium position, on the other hand, barring "permanent" changes in the environment, is a fixed population level (Griffiths, 1951; Strickland, 1954).

A species population is plastic and is undergoing constant change within the limits imposed upon it by its genetic constitution and the characteristics of its environment. Typical fluctuations in population and dispersion are shown in figure 3. The population dispersions shown at the three points in time A, B, and C are not static but rather are instantaneous phases of a continuously changing dispersion.

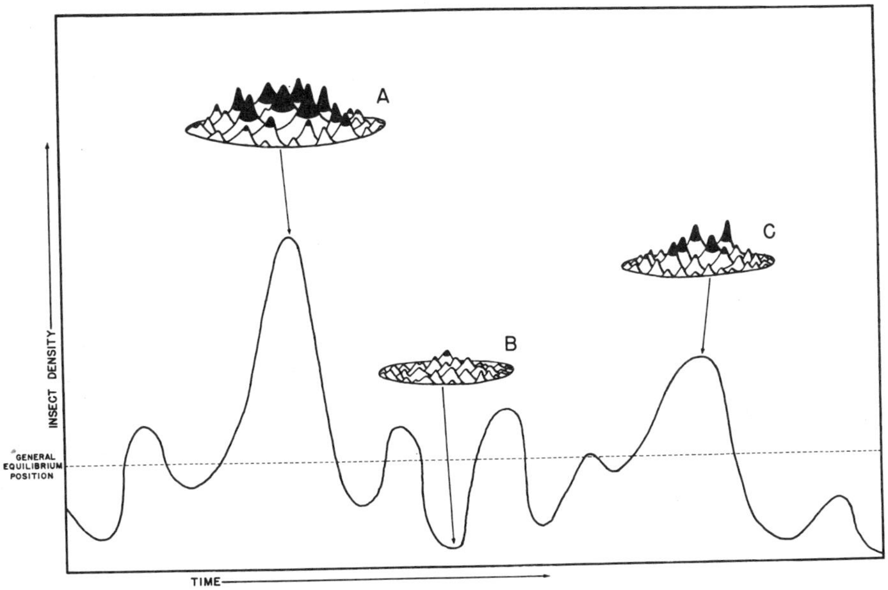

Fig. 3. Schematic graph of the population trend and population dispersion of a pest species over a long period of time. The solid line depicts the fluctuations in the population density with time. The broken line depicts the general equilibrium position. The population dispersion is indicated at the specific times *A, B,* and *C*. The basal area of these models reflects the distributional range, the height indicates population density. Population densities above the economic threshold are black.

Thus at point A, when the population is of greatest numerical abundance, it also has its widest distributional range (as depicted by the maximum diameter of the base of the model), and is of maximum economic status (as depicted by the number and magnitude of the blackened pinnacles representing penetrations of the economic threshold). At point B, on the other hand, when the species population is at its lowest numerical abundance, it is also most restricted in geographical range and is of only minor economic status. Point C represents an intermediate condition between points A and B.

In order to determine the relative economic importance of pest species, both the economic threshold and general equilibrium position of the pests must be considered. It is the general equilibrium position and its relation to the economic threshold, in conjunction with the frequency and amplitude of fluctuations about the general equilibrium position, that determine the severity of a particular pest problem.

In the absence of permanent modifications in the composition of the environment, the density of a species tends to fluctuate about the general equilibrium position as changes occur in the biotic and physical components of the environment. As the population density increases, the density-governing factors respond with greater and greater intensity to check the increase; as the population density decreases, these factors relax in their effects. The general equilibrium position is thus determined by the interaction of the species population, these density-governing factors, and the other natural factors of the environment. A permanent alteration of any factor of the environment—either physical or biotic—or the introduction of new factors may alter the general equilibrium position.

The economic threshold of a pest species can be at any level above or below the general equilibrium position or it can be at the same level. Some phytophagous species may utilize our crops as a food source but even at their highest attainable density are of little or no significance to man (see fig. 4, *A*). Such species can be found associated with nearly every crop of commercial concern.

Another group of arthropods rarely exceeds the economic thresholds and these consequently are occasional pests. Only at their highest population density will chemical control be necessary (see fig. 4, *B*).

When the general equilibrium position is close to the economic threshold, the population density will reach the threshold frequently (see fig. 4, *C*). In some cases, the general equilibrium position and the economic threshold are at essentially the same level. Thus, each time the population fluctuates up to the level of the general equilibrium position insecticidal treatment is necessary. In such species the frequency of chemical treatments is determined by the fluctuation rate about the general equilibrium position, which in some cases necessitates almost continuous treatment.

Finally, there are pest species in which the economic threshold lies below the general equilibrium position; these constitute the most severe pest problems in entomology (see fig. 4, *D*). The economic threshold may be lower than the level of the lowest population depression caused by the physical and biotic factors of the environment, *e.g.*, many insect vectors of viruses. In such cases, particularly where human health is concerned, there is a widespread and almost constant need for chemical control. This produces conditions favorable for development of insecticide resistance and other problems associated with heavy treatments.

One solution to pest problems and particularly those in this last category is to change the environment permanently so that the general equilibrium position will be lowered. For example, this might be accomplished through the introduction of a new biological control agent or through the permanent modification of a large portion of a required habitat. This has been done in certain areas with malaria-vector mosquitoes and similar pests by the draining of swamps and the destruction of other favorable habitats. Such methods may completely eliminate the species from some areas.

Environmental changes unfavorable to the pest may also be made through the use of plants and animals resistant to the pest species. This control method may involve three different aspects—tolerance, preference, and antibiosis (Painter, 1951). If tolerance alone is involved, the general equilibrium position may not be changed but the economic threshold is raised. Where preference or antibiosis is involved, the ability of the pest to reproduce upon the host is reduced, so that the general equilibrium position is lowered.

The lack of a sound measure of economic thresholds, in many cases, has been a major stumbling block to the development of integrated pest-control programs. Our changing economy, variations in natural governing mecha-

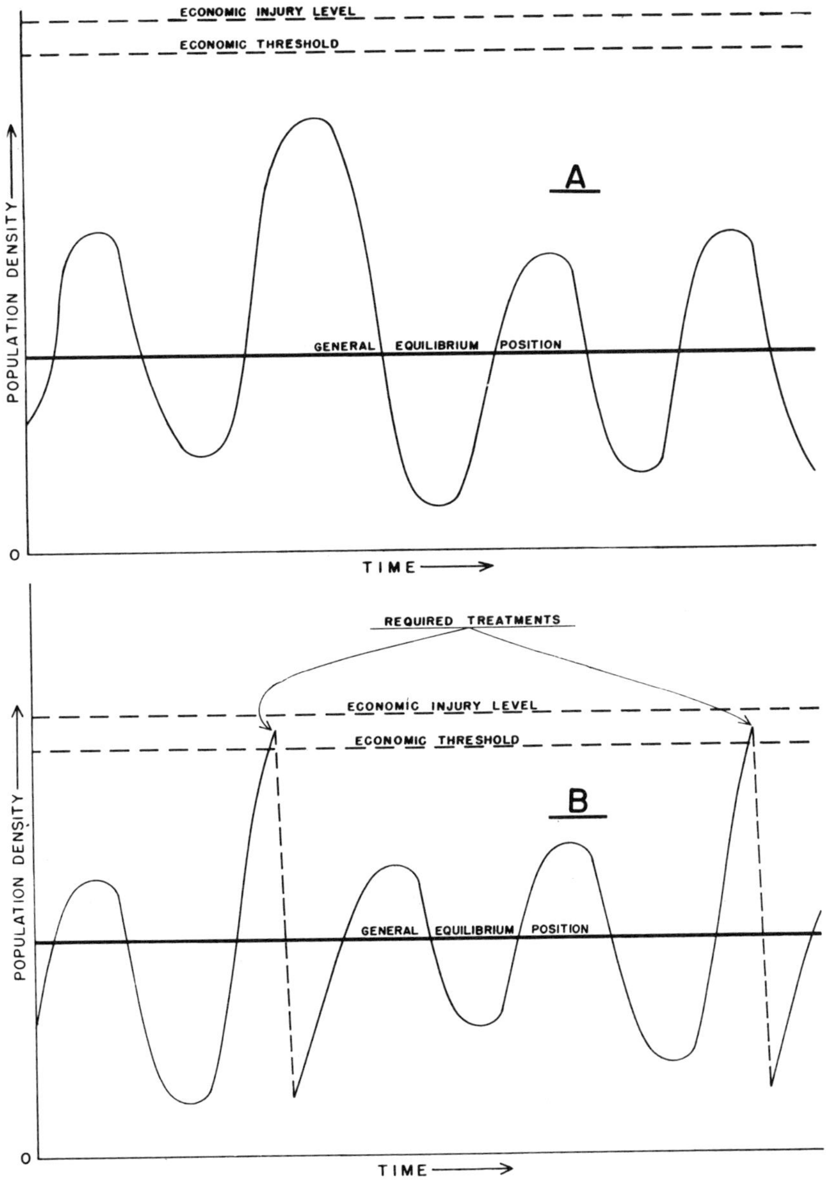

Fig. 4. Schematic graphs of the fluctuations of theoretical arthropod populations in relation to their general equilibrium position, economic thresholds, and economic-injury levels. *A*, Noneconomic population whose general equilibrium position and highest fluctuations are below the economic threshold, *e.g.*, *Aphis medicaginis* Koch on alfalfa in California. *B*, Occasional pest whose general equilibrium position is below the economic threshold but whose highest population fluctuations exceed the economic threshold, *e.g.*,

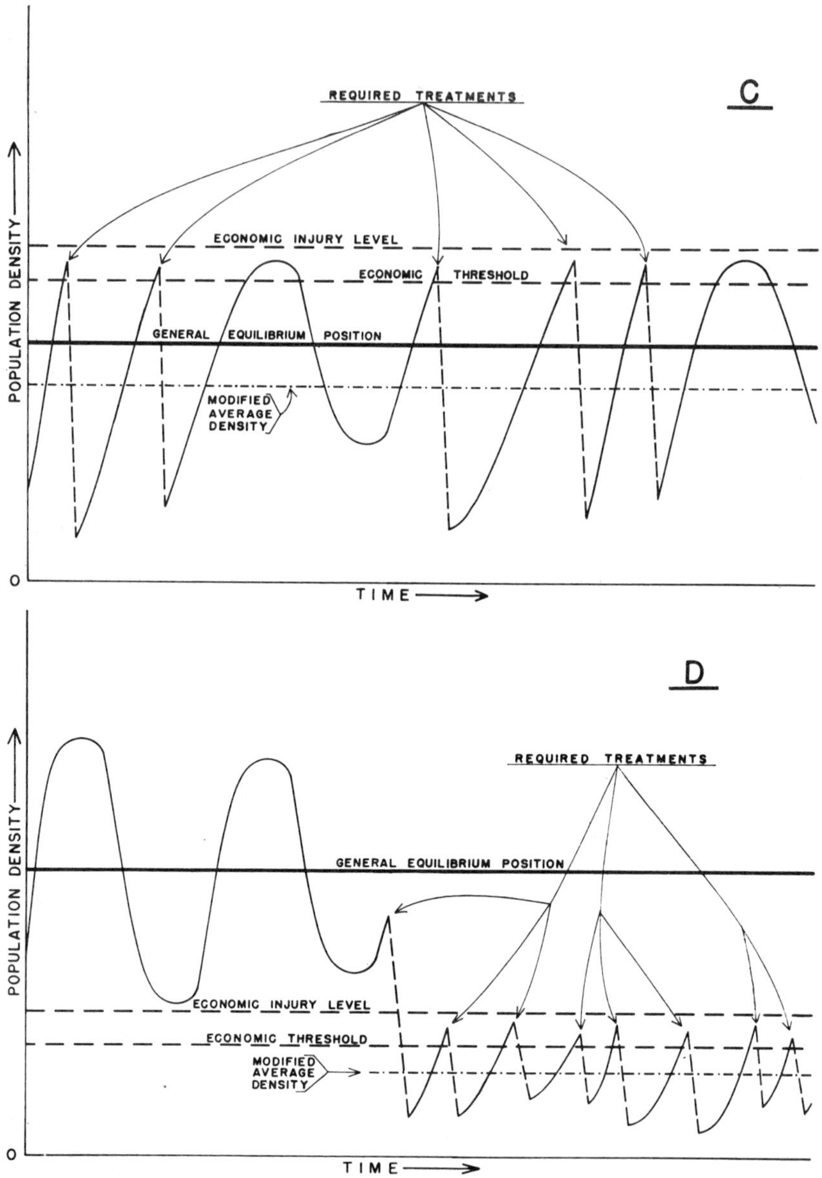

Grapholitha molesta Busck on peaches in California. *C,* Perennial pest whose general equilibrium position is below the economic threshold but whose population fluctuations frequently exceed the economic threshold, *e.g., Lygus* spp. on alfalfa seed in the western United States. *D,* Severe pest whose general equilibrium position is above the economic threshold and for which frequent and often widespread use of insecticides is required to prevent economic damage, *e.g., Musca domestica* in Grade A milking sheds.

nisms from one geographical area to another, differences in consumer demands, and the complexity of measuring the total effect of insects on yield and quality often make the assessment of economic damage extremely difficult. Yet the economic threshold and the economic-injury level must be determined reasonably and realistically before integrated pest control can develop to its fullest. Success in any well-balanced pest-control program is dependent on the aim of holding insect populations below experimentally established economic levels rather than attempting to eliminate all the insects.

THE INTEGRATION OF BIOLOGICAL AND CHEMICAL CONTROL

Biological control and chemical control are not necessarily alternative methods; in many cases they may be complementary, and, with adequate understanding, can be made to augment one another. One reason for the apparent incompatibility of biological and chemical control is our failure to recognize that the control of arthropod populations is a complex ecological problem. This leads to the error of imposing insecticides on the ecosystem, rather than fitting them into it. It is short-sighted to develop a chemical control program for the elimination of one insect pest and ignore the impact of that program on the other arthropods, both beneficial and harmful, in the ecosystem. On the other hand, this approach is no worse than the other extreme which would eliminate chemicals to preserve the biological control even in the face of serious economic damage. For we must recognize that modern agriculture could not exist without the use of insecticides. The evidence that biological and chemical control can be integrated is mounting. It has come from many sources involving many kinds of pests in various situations: see Ullyett (1947), Pickett and Patterson (1953), Ripper (1956), Huffaker and Kennett (1956), DeBach (1958*a*), Stern and van den Bosch (1959), and many others.

In approaching an integrated control program, we must realize that man has developed huge monocultures, he has eliminated forests and grasslands, selected special strains of plants and animals, moved them about, and in other ways altered the natural control that had developed over thousands upon thousands of years. We could not return to those original conditions if it were desirable. We may, however, utilize some of the mechanisms that existed before man's modifications, to establish new balances in our favor.

Recognition of the Ecosystem. To establish new, favorable balances, it is first necessary to recognize the "oneness" of any environment, natural or man-made. The populations of plants and animals (including man) and the nonliving environment together make up an integrated unit, the ecosystem. If an attempt is made to reduce the population level of one kind of animal (for example, a pest insect) by chemical treatment, modification of cultural practices, or by other means, other parts of the ecosystem will be affected as well. For this reason, the production of a given food or fiber must be considered in its entirety. This includes simultaneous consideration of insects, diseases, plant nutrition, plant physiology, and plant resistance, as well as the economics of the crops (Forbes, 1880; Ullyett, 1947; Pickett, 1949; DeBach, 1951; Solomon, 1953; Pickett and Patterson, 1953;

Glen, 1954; Michelbacher, 1954; Huffaker and Kennett, 1956; Simmonds, 1956; Balch, 1958; Decker, 1958; and others).

In most agricultural ecosystems, some potentially harmful organisms are continually held at subeconomic levels by natural controlling forces. In others, the pests are held below economic levels only part of the time. A pest species may be under satisfactory biological control over a large area or a long period of time, but not in all individual fields or during all periods. In a single field or orchard, or during a portion of a year, the pest population may rise to economic levels, while elsewhere or at other times the pest may be subeconomic. It is in such situations that integrated control programs are especially important. These intermittently destructive populations must be reduced in a manner that permits the biological control which prevailed before or prevails elsewhere to take over again. If a chemical treatment destroys the biotic agents without eradicating the pest, then repeated treatments may become necessary.

Population Sampling and Prediction. The sampling methods utilized by most research investigators for experimental plots are usually too time-consuming and tedious to be of practical value in establishing pest population levels in commercial crops. Special index methods are needed that are rapid and simple to use. Ideally, these should be of such nature that they can be easily utilized by the person examining the crop. But in many cases the grower is not able to evaluate all situations because of the difficulties and complexities involved in determining the status of some pest populations at the times of the year when they must be controlled. Then qualified entomologists will be required to evaluate the populations (Ripper, 1958).

One answer to this problem has been the development of supervised control in California, Arkansas, Arizona, and elsewhere. In a supervised control program the farmer, or a group of farmers, contracts with a professional entomologist who determines the status of the insect populations. On the basis of his population counts, other conditions peculiar to each situation, and his knowledge of the ecology of the pests and their biological controls, the entomologist makes predictions as to the course of the population trends and advises as to when controls should be applied and what kind. For instance, in the case of the alfalfa caterpillar, *Colias philodice eurytheme*, when economic thresholds are reached, the recommended procedure may involve immediate cutting of the hay crop without treatment, application of the polyhedrosis virus (Steinhaus and Thompson, 1949; Thompson and Steinhaus, 1950), or treatment with an insecticide. The course to be taken depends on the characteristics of the particular infestation (Smith and Allen, 1954).

Wherever possible, knowledge must be developed so that we can predict the times when occasional pests will be present in outbreak numbers. This will eliminate unnecessary and environment-disturbing "insurance" treatments. When this is not possible, the treatments can be timed according to the actual pest population levels, as is now done with many field-crop pests.

With those crops that do not yet have fixed chemical control schedules, every effort should be made to plan programs dependent upon pest population levels and to avoid dependence upon insurance and prophylactic treatments. If this is not done there is real danger that on these crops, too, pest-control problems will become increasingly complex.

Augmentation of Natural Enemies. In some situations, the development of integrated control requires the augmentation of the natural-enemy complex (DeBach, 1958*a*). The introduction of additional natural enemies is usually the simplest and best solution. This may not be possible or effective with some pest species, and methods of overcoming the inefficiency of the natural-enemy complex must be sought. This can be done by periodic colonization of parasite or predators (Doutt and Hagen, 1950; DeBach, Landi, and White, 1955), artificial inoculation of the host at times of low density (Smith and DeBach, 1953; Huffaker and Kennett, 1956), modification of the environment, or selective breeding of parasites and predators.

The modification of the environment may involve changes in irrigation, introduction of a covercrop, or development of greater plant heterogeneity. Refuges for beneficial forms can be produced by strip treatments with chemicals (DeBach, 1958*a*) or by the development of uncultivated and untreated areas (Grison and Biliotti, 1953). Modification of the environment may also involve the control of ants or other organisms which curtail parasite and predator activity (Flanders, 1945; DeBach, Fleschner, and Dietrick, 1951). The selective breeding of parasites and predators may be directed toward increased or modified tolerance ranges to physical conditions (DeBach, 1958*b*) or insecticides (Robertson, 1957).

Where prophylactic treatments are proved to be necessary for a perennial pest, selective materials must be developed and utilized to foster biological control both of other pests and of the pest of direct concern at other times.

Selective Insecticides. Chemical control programs are limited by the nature of the available materials. In the past, nonselective insecticides applied for one insect in a pest complex often have eliminated the biotic factors holding other pests in check. More recently, we have had available a greater variety of materials, some of them selective in their action (Ripper, 1956).

The selective use of insecticides may be accomplished in at least four ways. First, the insecticide itself may be selective in its toxicological action. Narrow-range toxicants may be utilized to reduce a pest of concern and at the same time spare the beneficial forms (Ripper, 1944; Ripper, Greenslade, and Hartley, 1951). A particular material may be selective in one situation and not in another; or it may be selective at low dosages but not at high dosages. Furthermore, the manner of application (Ripper, 1956) and especially the type of carrier and residue deposit may produce differential effects on the insect complex (Flanders, 1941; Holloway and Young, 1943; DeBach and Bartlett, 1951).

Second, we can produce a selective action on a pest-parasite complex by treating only those areas where the pest-parasite ratio is unfavorable. This method is one of the bases of supervised control of the alfalfa caterpillar in California (Smith and Allen, 1954). Population levels of both the host caterpillar and its parasite, *Apanteles medicaginis* Muesebeck, are determined at appropriate intervals in all fields. A prediction of possible damage is made on the basis of these population levels, and only those infestations which are potentially damaging are treated. In this way, on an area-wide basis, the balance is shifted in favor of the parasites, even though parasite adults and parasite larvae within the host caterpillars are often destroyed in the treated fields. The success of such programs will depend on the exact nature of the local problem and the quality of supervision. The rates of dispersal of parasites, predators, and pests are complicating factors.

Third, proper timing of insecticides can produce a selective action on the pest and natural-enemy complex (Ewart and DeBach, 1947; Michelbacher and Middlekauff, 1950; Bartlett and Ewart, 1951; Jeppson, Jesser, and Complin, 1953; Massee, 1954; DeBach, 1955). In such situations, an intimate knowledge of the behavior patterns of the pests and their natural enemies is required.

Fourth, nonselective materials with short residual action may be used if the beneficial forms can survive in a resistant stage or in an untreated reservoir area. Stern and van den Bosch (1959) have demonstrated that parasites of the spotted alfalfa aphid can survive nonselective treatments if they are in the more resistant pupal stage. DeBach (1958*a*) reports successful integration of biological and chemical control of purple scale on citrus where alternate pairs of tree rows were sprayed at 6-month intervals with a nonselective oil treatment.

For some pests a disease pathogen may be used as a selective insecticide (Steinhaus, 1954; 1956). For example, under supervised control in the Dos Palos area of California, the polyhedrosis virus affecting the alfalfa caterpillar has been used successfully either alone or in combination with a selective insecticide to avoid the use of a nonselective treatment. More recently, interest has developed in the commercial use of virulent strains of *Bacillus thuringiensis* Berliner, for the control of certain truck- and field-crop pests in California. The use of disease pathogens as selective insecticides is in its infancy, but can be expected to increase in importance with additional research (Steinhaus, 1951, 1957).

The ideal selective material is not one that eliminates all individuals of the pest species while leaving all of the natural enemies. Use of such a material would force the predators and parasites to leave the treated area or starve (Clausen, 1936; Flanders, 1940). The ideal material is one that shifts the balance back in favor of the natural enemies (Boyce, 1936; Ripper, 1944; Wigglesworth, 1945).

Future of Integrated Control. If our knowledge were adequate today to outline an ideal integrated control program for a crop now utilizing an intensive fixed spray program, it would not be possible to switch to such a program immediately. The effects of the previous treatments may last several years. In some instances, effective biological control no longer exists and would have to be reëstablished. This may be a slow process (DeBach, 1951; DeBach and Bartlett, 1951; Pickett and Patterson, 1953; and others).

It should be emphasized also that the development of integrated control is not a panacea that can be applied blindly to all situations, for it will not work if biotic mortality agents are inadequate or if low economic thresholds preclude utilizing biological control (Barnes, 1959). However, it has worked so well in some appropriate situations that there can be no doubt as to its enormous advantages and its promise for the future.

LITERATURE CITED

BALCH, R. E.
1958. Control of forest insects. Ann. Rev. Ent. 3:449–68.

BARNES, M. M.
1959. Deciduous fruit insects and their control. Ann. Rev. Ent. 4:343–62.

BARTLETT, B., and W. H. EWART
1951. Effect of parathion on parasites of *Coccus hesperidum*. Jour. Econ. Ent. **44**: 344–347.

BARTLETT, B. R., and J. C. ORTEGA
1952. Relation between natural enemies and DDT-induced increases in frosted scale and other pests of walnuts. Jour. Econ. Ent. **45**(5):783–85.

BOVEY, PAUL
1955. Les actions secondaires des traitements antiparasitaires sur les populations d'insectes et d'acariens nuisibles. Schweiz. Landw. Monatsh. **33** (9/10):369–79.

BOYCE, A. M.
1936. The citrus red mite, *Paratetranychus citri* McG. in California, and its control. Jour. Econ. Ent. **29** (1):125–30.

BROWN, A. W. A.
1951. Insect control by chemicals. John Wiley & Sons, Inc., New York, N.Y. 817 pp.
1959. Spread of insecticide resistance. Adv. Pest Control Res. **2**:351–414. Interscience Publishers, Inc., New York, N.Y.

CLAUSEN, C. P.
1936. Insect parasitism and biological control. Ent. Soc. Amer. Ann. **29**:201–23.
1956. Biological control of insect pests in the continental United States. U.S. Dept. Agr. Tech. Bul. 1139. 151 pp.

CROW, J. F.
1957. Genetics of insect resistance to chemicals. Ann. Rev. Ent. **2**:227–46.

DEBACH, P.
1951. The necessity for an ecological approach to pest control on citrus in California. Jour. Econ. Ent. **44**(4):443–47.
1955. Validity of the insecticidal check method as a measure of the effectiveness of natural enemies of diaspine scale insects. Jour. Econ. Ent. **48**(5):584–88.
1958*a*. Application of ecological information to control of citrus pests in California. Tenth Internatl. Congr. Ent. Proc. **3**:185–97.
1958*b*. Selective breeding to improve adaptations of parasitic insects. Tenth Internatl. Congr. Ent. Proc. **4**:759–67.

DEBACH, P., and B. BARTLETT
1951. Effects of insecticides on biological control of insect pests of citrus. Jour. Econ. Ent. **44**(3):372–83.

DEBACH, P., C. A. FLESCHNER, and E. J. DIETRICK
1951. A biological check method for evaluating the effectiveness of entomophagous insects. Jour. Econ. Ent. **44**(5):763–66.

DEBACH, P., J. H. LANDI, and E. B. WHITE
1955. Biological control of red scale. California Citrog. **40**:254, 271–72, 274–75.

DECKER, G. C.
1958. Don't let the insects rule. Agr. Food Chem. **6**(2):98–103.

DOUTT, R. L.
1958. Vice, virtue and the vedalia. Ent. Soc. Amer. Bul. **4**(4):119–23.

DOUTT, R. L., and K. S. HAGEN
1950. Biological control measures applied against *Pseudococcus maritimus* on pears. Jour. Econ. Ent. **43** (1):94–96.

ELTON, CHARLES S.
1958. The ecology of invasions by animals and plants. 181 pp. Methuen & Co. Ltd., London; John Wiley & Sons Inc., New York, N.Y.

EWART, W. H., and P. DEBACH
1947. DDT for control of citrus thrips and citricola scale. California Citrog. **32**:242–45.
FLANDERS, S. E.
1940. Environmental resistance to the establishment of parasitic Hymenoptera. Ent. Soc. Amer. Ann., **33**:245–53.
1941. Dust as an inhibiting factor in the reproduction of insects. Jour. Econ. Ent. **34**(3):470–1.
1945. Coincident infestations of *Aonidiella atrana* and *Coccus hesperidum*, a result of ant activity. Jour. Econ. Ent. **38**:711–12.
FLESCHNER, C. A., and G. T. SCRIVEN
1957. Effects of soil-type and DDT on ovipositional response of *Chrysopa californica* (Coq.) on lemon trees. Jour. Econ. Ent. **50**(2):221–22.
FORBES, S. A.
1880. On some interactions of organisms. Illinois State Lab. Nat. Hist. Bul. **3**:3–17.
GLEN, R.
1954. Factors that affect insect abundance. Jour. Econ. Ent. **47**:398–405.
GRIFFITHS, J. T.
1951. Possibilities for better citrus insect control through the study of the ecological effects of spray programs. Jour. Econ. Ent. **44**:464–68.
GRISON, P., and E. BILIOTTI
1953. La signification agricole des "stations refuges" pour la faune entomologique. Acad. d'Agr. de France Compt. Rend. **39**(2):106–9.
HAYES, W. J.
1954. Agricultural chemicals and public health. U. S. Public Health Reports **69**(10): 893–98.
HERTZLER, J. O.
1956. The crisis in world population. 279 pp. Univ. Nebraska Press, Lincoln.
HOLLOWAY, J. K., and T. ROY YOUNG, JR.
1943. The influence of fungicidal sprays on entomogenous fungi and on the purple scale in Florida. Jour. Econ. Ent. **36**(3):453–57.
HOSKINS, W. M., and H. T. GORDON
1956. Arthropod resistance to chemicals. Ann. Rev. Ent. **1**:89–148.
HUFFAKER, C. B., and C. E. KENNETT
1956. Experimental studies on predation: Predation and cyclamen-mite populations on strawberries in California. Hilgardia **26**(4):191–222.
JEPPSON, L. R., M. J. JESSER, and J. O. COMPLIN
1953. Timing of treatments for control of citrus red mite on orange trees in coastal districts of California. Jour. Econ. Ent. **46**:10–14.
KLOSTERMEYER, E. C., and W. B. RASMUSSEN
1953. The effect of soil insecticide treatments on mite populations and damage. Jour. Econ. Ent. **46**:910–12.
LINSLEY, E. G., editor
1956. Evaluation of certain acaricides and insecticides for effectiveness, residues, and influence on crop flavor. Hilgardia **26**(1):1–106.
LORD, F. T.
1947. The influence of spray programs on the fauna of apple orchards in Nova Scotia: II. Oystershell scale, *Lepidosaphis ulmi* (L.) Canad. Ent. **79**:196–209.
MASSEE, A. M.
1954. Problems arising from the use of insecticides: effect on the balance of animal populations. 6th Commonwealth Entomol. Conf. Rept. Pp. 53–57. London, England.
METCALF, R. L.
1955. Physiological basis for insect resistance. Physiol. Rev. **35**:197–232.
MICHELBACHER, A. E.
1954. Natural control of insect pests. Jour. Econ. Ent. **47**(1):192–94.

MICHELBACHER, A. E., and S. HITCHCOCK

1958. Induced increase of soft scales on walnut. Jour. Econ. Ent. **51**(4):427–31.

MICHELBACHER, A. E., and W. W. MIDDLEKAUFF

1950. Control of the melon aphid in northern California. Jour. Econ. Ent. **43**(4): 444–47.

PAINTER, R. H.

1951. Insect resistance in crop plants. xi + 520 pp. Illus. Macmillan Company, New York, N.Y.

PETTY, C. S.

1957. Organic phosphate insecticide poisoning; an agricultural occupational hazard. Louisiana St. Med. Soc. Jour. **109**(5):158–64.

PICKETT, A. D.

1949. A critique on insect chemical control methods. Canad. Ent. **81**(3):67–76.

PICKETT, A. D., and N. A. PATTERSON

1953. The influence of spray programs on the fauna of apple orchards in Nova Scotia. IV. A Review. Canad. Ent. **85**(12):472–78.

PICKETT, A. D., W. L. PUTMAN, and E. J. LE ROUX

1958. Progress in harmonizing biological and chemical control of orchard pests in eastern Canada. Tenth Internatl. Congr. Ent. Proc. **3**:169–74.

RIPPER, W. E.

1944. Biological control as a supplement to chemical control of insect pests. Nature **153**:448–52.

1956. Effect of pesticides on balance of arthropod populations. Ann. Rev. Ent. **1**:403–38.

1958. The place of contracting organizations and professional supervision in the application of pest control methods. Tenth Internatl. Congr. Ent. Proc. **3**:93–97.

RIPPER, W. E., R. M. GREENSLADE, and G. S. HARTLEY

1951. Selective insecticides and biological control. Jour. Econ. Ent. **44**(4):448–59.

ROBERTSON, J. G.

1957. Changes in resistance to DDT in *Macrocentrus ancylivorus* Rohw. Canad. Jour. Zool. **35**(5):629–633.

SABROSKY, C. W.

1952. How many insects are there? Pp. 1–7 *in:* Insects, 1952 Yearbook of Agriculture. 780 pp. U. S. Dept. Agr., Washington, D.C.

SCHNEIDER, F.

1955. Beziehungen zwischen nützlingen und chemischer Schädlingsbekämpfung. Deutsche Gesell. f. Angw. Ent. Verhandl. (13te Mitglied versamm.). **1955**:18–29.

SIMMONDS, F. J.

1956. The present status of biological control. Canad. Ent. **88**(9):553–63.

SMITH, H. S., and P. DEBACH

1953. Artificial infestation of plants with pest insects as an aid in biological control. Seventh Pacific Sci. Congr. Zoology (1949) Proc. **4**:255–59.

SMITH, RAY F.

1959. The spread of the spotted alfalfa aphid, *Therioaphis maculata* (Buckton), in California. Hilgardia **28**(21):647–91.

SMITH, R. F., and W. W. ALLEN

1954. Insect control and the balance of nature. Sci. Amer. **190**(6):38–42.

SOLOMON, M. E.

1953. Insect population balance and chemical control of pests. Chem. and Indus. **1953**: 1143–47.

STEINHAUS, E. A.

1951. Possible use of *Bacillus thuringiensis* Berliner as an aid in the biological control of the alfalfa caterpillar. Hilgardia **20**(18):359–81.

1954. The effects of disease on insect populations. Hilgardia **23**(9):197–261.

1956. Potentialities for microbial control of insects. Agr. Food Chem. **4**(8):676–80.

1957. Concerning the harmlessness of insect pathogens and the standardization of microbial control products. Jour. Econ. Ent. **50**(6): 715–20.

STEINHAUS, E. A., and C. G. THOMPSON
1949. Preliminary field tests using a polyhedrosis virus in the control of the alfalfa caterpillar. Jour. Econ. Ent. **42**(2):301–5.
STERN, V. M., and R. VAN DEN BOSCH
1959. Field experiments on the effects of insecticides. Hilgardia **29**(2):103–30.
STRICKLAND, A. H.
1954. The assessment of insect pest density in relation to crop losses. 6th Commonwealth Ent. Conf. Rept., pp. 78–83. London, England.
THOMAS, W. L., JR., EDITOR
1956. Man's role in changing the face of the earth. 1193 pp. Univ. Chicago Press, Chicago, Ill.
THOMPSON, C. G., and E. A. STEINHAUS
1950. Further tests using a polyhedrosis virus to control the alfalfa caterpillar. Hilgardia **19**(14):411–45.
TROUVELOT, B.
1936. Le doryphore de la pomme de terre (*Leptinotarsa decemlineata* Say) en Amerique de Nord. Ann. Epiphyt. (n.s.) **1**:277–336.
ULLYETT, G. C.
1947. Mortality factors in populations of *Plutella maculipennis* (Tineidae: Lep.) and their relation to the problem of control. South African Dept. Agr. and Forestry Ent. Mem. **2**(6):77–202.
1951. Insects, man and the environment. Jour. Econ. Ent. **44**(4):459–64.
UPHOLT, W. M.
1955. Evaluating hazards in pesticides use. Agr. Food Chem. 3(12):1000–6.
WILLE, J. E.
1951. Biological control of certain cotton insects and the application of new organic insecticides in Peru. Jour. Econ. Ent. **44**:13–18.
WIGGLESWORTH, V. B.
1945. DDT and the balance of nature. Atlantic Monthly **176**(6):107–13.

PESTICIDES

PESTICIDES: THEIR USE AND LIMITATIONS IN PEST MANAGEMENT

RAY F. SMITH

Pesticides remain, in spite of their adverse publicity, our most powerful tool in the management of pests. The pesticides, as developed in the last 25 years, are highly effective and economical and can be marshalled quickly to have immediate impact on a pest population—even over a large area. When pest populations approach economic levels, there is little other than pesticides we can use to avoid damage and which will have the desired immediate effect. Hence, it seems clear that pesticides must and will continue to be used in a major way in pest management. In fact, few informed and rational people desire or anticipate the elimination of all pesticides.

Furthermore, the "Green Revolution" which is occurring in many countries through the introduction of new high-yielding varieties of wheat, rice, maize, sorghum and other food grains has brought on "second-generation" problems (Brown 1970). These might be termed problems of success, and one of them is the intensification of pest problems. The basis for this pest increase is multifactorial. The precise details differ with the crop variety involved and local conditions prevailing in the country. For example, the variety may be more suitable for a pest. This is the case with IR-8 rice which is an improved host for brown plant hopper as compared to old varieties. The combination of better water management and increased fertilization as is taking place in the "Green Revolution" may make for higher yields of grain, but it may also produce larger insects which live longer and lay more eggs. The new cropping conditions may also create a microenvironment favoring new insect pests and pathogens or increased levels of old ones. The higher yields also increase the farmer's income substantially and permit him to use more pesticides. These are some of the factors which are contributing to the increased use of pesticides, especially in the less developed countries. Let us hope that this increased use will come about in such a manner as to avoid the serious consequences which have come with the intensive use of pesticides in the developed countries.

The disadvantages or limitations of pesticide chemicals are well-known. They have been so emphasized in the press, on radio and TV, in political arenas, and elsewhere that it is now difficult to have a rational discussion that balances the good and the undesirable aspects of pesticides. In brief review, the limitations of pesticide usage can be listed as follows:

a. Development of strains of pests that are resistant to pesticides.

b. Temporary effects on pest populations necessitating repeated treatment. (Often the pest population quickly returns to a higher level than before treatment.)

c. Hazards from residues of the pesticide in the harvested crop.

CONCEPTS OF PEST MANAGEMENT, Proceedings of Conference, 1970, vol. 1, pp. 103-113.

d. Outbreaks of unleashed secondary pests resulting from the destruction of their natural enemies.

e. Undesirable side-effects on non-target organisms including 1) parasites and predators; 2) fish, birds, and other wildlife; 3) honey bees and other necessary pollinators; 4) man and his domestic animals; and 5) the crop plant.

f. Direct hazards in the application of pesticides.

g. Reduction and simplification of the biotic component of the agro-ecosystem.

This is a formidable list that should make any wise person pause in his continued use of pesticides. In a more positive way, I prefer to look upon the above list as a basic guide to the better use of pesticides for pest management. If we can devise procedures for the use of pesticides which will avoid or minimize these disadvantages or complications, then we will have developed a methodology for the efficient and effective use of pesticides. But how? Let us look more closely at these disadvantages.

INSECT RESISTANCE

The development of selected strains of insects which can no longer be controlled by an insecticide at the rates normally recommended (or often at any rate) has become one of the most serious limitations in the use of pesticides. The development of resistance to DDT and other chlorinated hydrocarbon insecticides has spelled out the demise of these materials far more effectively than any laws or regulations banning or restricting their use. Well over two hundred species of insects, mites and ticks are now documented as having developed resistance to one or more pesticides (Brown 1968, FAO 1969). It is difficult to estimate what proportion of the insect pests of the world subjected to pesticide selective pressure have developed resistance, but we are sure that the proportion is large. In fact, only a very few species regularly treated with pesticides have not developed significant levels of resistance. The European cornborer, *Ostrinia nubilalis*, is frequently cited as an example of a species which has not developed resistance to DDT. However, the European cornborer has but one or only a few generations per year and only a small portion of the population is treated each year. Hence, I would question if the European cornborer has been subjected to a severe selection pressure. In other examples, the selection pressure from pesticides is reduced by the fact the pest species has many hosts (only one or a few of which is treated), or it occurs in large areas of uncultivated land (hence a large proportion of the population escapes treatment), or it is very vagile (thus the treated and untreated parts of the population mix rapidly). Exceptions to the development of resistance under heavy selection pressure should be scrutinized most carefully for what we might learn to avoid the development of resistance. In any case, as a precautionary planning measure we must assume that any arthropod exposed to intensive pesticide pressure will develop insecticide resistance.

What should be our strategy in the use of pesticides to avoid or delay the development of resistance? It seems to me the way is clear and straightforward although it is not at all an easy one. We must minimize the selection pressure on the pest populations hopefully to a level that other countering selection pressures will offset any trend toward resistance (Waterhouse 1969). This can be done by avoiding all but the absolutely necessary treatments and utilizing as many alternative control procedures as possible so that we will not be entirely dependent on pesticides. This, of course, is another definition for integrated control.

The development of resistance in any particular pest population depends first on whether genes for resistance are present in the population. Experience indicates that such genes are usually present. Secondly, the intensity of the selection pressure for resistance is most important. We can have considerable control over this selection pressure and we should take advantage of it. Thirdly, the characteristics of species and populations such as dispersal behavior, generation time, reproductive rate, and degree of isolation determines the pattern and rate of development of resistance. In integrated control systems, there is usually less risk of resistance as compared with programs involving complete reliance on use of pesticides. This results from lowered selection pressures because pesticides are no longer responsible for virtually all of the mortality. For example, in integrated control systems natural enemies are protected and utilized as fully as possible and pesticides are applied only to those areas with economic populations of pests (Smith and van den Bosch 1967). In other words, the selective action of the pesticide usage is directed against only relatively small portions of the population scattered in time and space. In the intervals between pesticide treatments, different selective pressures from other control procedures and from other elements of the environment will modify the large remaining population (in untreated cultivated areas and in uncultivated areas) and may counteract trends toward resistance. The use of selective pesticide materials enhances this pattern and also avoids the development of resistance in other pest species associated with the target species.

It is now an essential part of modern pest control procedures to monitor continuously the level of pesticide resistance in our major pests. Not only will this tell us when a particular pesticide is starting to lose its usefulness, but it also will give us a check on other procedures which may delay or prevent the development of resistance. For example, the selection of strains of mites resistant to acaricides has been slowed in Washington State apple orchards where integrated control practice has reduced the selection pressure from pesticides (Hoyt 1970).

Two factors have been of major importance in avoiding unnecessary pesticide treatments in integrated control systems. The first is utilization of economic injury levels in determining need for pesticide treatment and the second is supervised control for delimiting the precise areas requiring treatment. The necessary chemical pesticide treatments should be designed to reduce pest abundance so that economic damage is prevented. This requires the establishment of sound economic injury levels (Smith 1969b). This is not easy and ideally will involve the combined talents of both pest management specialists and agricultural economists. Initially in an integrated control system, the determination of economic injury levels can be generally based on empirical evidence—that is, past experience with the pests. Later, however, these levels should be reviewed constantly and retested in accordance with changes in farming practice and, as additional information is obtained, from further observations and from carefully controlled experiments. The ultimate will be to have rather elaborate economic injury levels qualified in terms of local climate, time of year, stage of plant development, plant variety, cropping practices, and economic variables.

The selective treatment of local areas of pest abundance becomes especially important in the avoidance of pesticide resistance because this procedure leaves untreated and hence unselected large components of the total gene pool. Supervised control systems can be used for this purpose and such systems have been utilized for a large variety of crops. My own experience involves programs developed for alfalfa and cotton. A supervising entomologist or his assistants periodically determine the densities of pests and beneficial species throughout the cropped area. On the basis of this information and a knowledge of the organisms involved, recommendations are made to the farmers as to which areas should be treated and which not.

Fixed treatment schedules, such as treatments by the calendar, are almost certain to result in many unnecessary applications of pesticides thus intensifying the resistance problem. They give natural enemies no refuge or chance for recovery. Consequently, such rigid treatment schedules have frequently engendered severe resurgence of target pests and rise of secondary pests, both conditions requiring still additional treatments and intensification of selection for resistance.

In general, integrated control systems tend to be more sophisticated and complicated in their application than those which rely solely on pesticides. For this reason, too, it has been frequently necessary to develop supervised control to monitor and refine the day to day control procedures. This in itself has been a limitation of use for pesticides because of the limited supply of qualified supervising entomologists.

A most valuable adjunct to the scientific use of pesticides is a system for accurately predicting or determining damaging population levels. This permits the invoking of additional control efforts at critical times and their withdrawal when the crises have passed. Such a system has the necessary flexibility to adjust to changing seasons, cultural patterns, weather variations. etc. The prediction system permits the pesticides to be used selectively and only on potentially economically threatening populations. In recent years, light traps have been widely used to assist population prediction (van den Bosch et al 1970). A supervised control system is greatly strengthened in providing economical pest control and is more effective in avoiding or slowing development of resistance when coupled with a sound pest outbreak prediction system.

TEMPORARY CONTROL EFFECT OF PESTICIDES

In contrast to some of the living controlling factors such as parasites and predators, chemical pesticides are non-reproductive, have no searching capacity, and are not self-sustaining. This is one of the inherent limitations of pesticides that we must learn to live with: If we use higher dosages of persistent pesticides to off-set this deficiency, then we run into aggravated problems with pesticide residues, secondary outbreaks, and resistance tends to develop more rapidly.

Moreover, the resurgence of the pest population is greatly accelerated in some such circumstances. For example, where a broad-spectrum pesticide is used (and natural enemies of the pest are also eliminated) or when high dosages of a selective material are used that give very high percentage kills of the pest (and thus because their food is eliminated the natural enemies are starved out), the pest population can recover quickly without the hindrance of natural enemies. Pesticides when used in integrated control systems should aim to reduce the pest population below the economic threshold level without severely disturbing the natural enemies. Under these circumstances, as has been exemplified in the integrated control of the spotted alfalfa aphid, the natural enemies can take over and maintain the level of control initiated by the selective chemical. Such successes depend on the selective use of pesticides and confidence in the fact that we do not need full elimination of the pest. We have encountered our greatest troubles with pesticides when we have tried to achieve full control with pesticides alone.

RESIDUE HAZARDS

Pesticide chemicals may remain in or on the crop plants for extended periods after application. They may drift to other nearby crop areas or drain into streams and other waterways. They may thereby create a hazard to man or domesticated and wild animals and in other ways cause unwanted side effects.

The residue problem will always confront us when we use toxic chemicals with any significant degree of persistence. The residue problem can be greatly diminished through proper selection of an appropriate pesticide and its formulation and by applying the material with suitable techniques at appropriate dosages. This is a matter to be solved by the economic entomologist and much of his field research must be devoted to such considerations.

In the past year or more there has been considerable public pressure to eliminate all persistent pesticides. In our response to this pressure let us not forget that society is losing valuable tools for pest control and that there are other problems associated with many of the substitute materials. To date, the substitutes are usually either organophosphorus materials or carbamates. It is difficult to generalize because there are so many exceptions. Basically each compound should be considered individually as to its peculiar risks on the basis of its specific characteristics together with the exact manner of dosage and place of application.

Nevertheless, of primary significance is a general pattern of higher acute toxicity in the substitute material with greater direct risk to man, livestock, and wildlife. Secondly, these substitute compounds frequently have produced serious damage to honeybees and other necessary pollinators. The impact of these substitute chemicals on the California beekeeping industry has been disastrous (Swift 1970). Thirdly, they have a severe impact on insect natural enemies, and the resultant elimination of these forms from treated areas frequently permits rapid resurgence of the target pests and outbreaks of previously innocuous species. Finally, the short-lived nature of the substitute materials together with this side-effect on natural enemies requires repetition of applications. This increases the selection pressure for resistance and hastens the development of populations resistant to the chemical.

Persistence of a pesticide chemical is not in itself an obnoxious quality. We need some level of persistence. This is especially true when the pest population moves slowly out of hibernation quarters or other inaccessible or untreatable habitats into the area of contact with the pesticide or into a susceptible stage of development. If the movement into the area to be protected extends over an appreciable length of time the pesticide must persist over this length of time or repeated treatments of a non-persistent material are necessary. In general, the latter procedure will be more costly and more hazardous. Persistence is a disadvantage when it is the cause of undesirable residues on the harvested crop or elsewhere in the agroecosystem or the pesticide is concentrated through food chains to harmful levels in non-target organisms. Again we must strike a balance among costs of alternative procedures and among the benefits and undesirable effects of these procedures. This comparison should not be made only in narrow economic terms but also with full consideration of the social costs and benefits.

DESTRUCTION OF NATURAL ENEMIES

As commented on previously, the destruction of natural enemies as an unfortunate side-effect of pesticide usage has two main consequences. First, the target pest may quickly recover from the pesticide action and resurge to even higher levels. Secondly, the resurgence of unleashed secondary pests may occur shortly after the application of the triggering pesticide, or later in the growing season, or even in a subsequent season. In cotton in California, for example, we have had serious outbreaks of such unleashed secondary pests as beet armyworm, cotton leaf perforator and cabbage looper. These secondary pests may be more destructive and more difficult to control than the original target pest. The re-establishment of the effectiveness of natural enemies may require two or more years (Flaherty and Huffaker, unpublished data).

The phenomena have been so well documented in recent years that it does not seem necessary to cite extensive examples here. To refresh your memory I will, however, mention the severe and destructive outbreaks of the cotton leaf perforator and to a lesser extent spider mites that occurred in the Imperial Valley of California as the result of a heavy chemical treatment schedule for cotton to "eradicate" pink bollworm. Natural enemies of many pests were eliminated and later when the cotton treatments were stopped, the most severe outbreak ever recorded in this area of cotton loopers and beet armyworms deleveloped on a variety of fall-planted crops. However, this does not mean all pest outbreaks result from unwise use of pesticides. The sudden outbreak in 1964 of a formerly-rare web-spinning sawfly, *Pamphilus phyllisae*, on northern red oak in Pennsylvania was probably brought on by a period of abnormally low precipitation (Eidt and Nichols 1970). Many such weather-released pest outbreaks occur.

Of great significance to our considerations of pest management is the question, "What can be done about secondary pest outbreaks?" The answer lies in the selective use of pesticides. All pesticides have some selectivity, but the range in degree of selectivity is very great. Much effort has been expended in seeking materials with relatively high toxicity to invertebrates and low toxicity to mammals. We must also seek differential toxicity within the Arthropoda. This is a difficult goal, but we do not need the ultimate in specificity that would permit us to prescribe a specific chemical for each pest species. However, we do need materials that are specific for groups of pests such as aphids, locusts, lepidopterous larvae, weevils, and so forth. There is no question but that we will have to pay more for such materials than we are now paying for broad-spectrum multiple-use materials available now. But there are indications, e.g., with certain acaricides and aphicides, that we can afford selective materials and that they will be economically feasible from the viewpoint of the chemical pesticide industry.

The above comments concern differential physiological effects of pesticides on different types of organisms. While we await the development of these materials we should enhance selectivity of available ones by taking advantage of ecological and behavioral characteristics of the pest species as well as the physical characteristics of formulations and of application techniques to obtain differential kill of desirable and undesirable organisms (i.e., selective action). The difference in the nocturnal resting sites of the house fly and other flies (walls and ceilings of chicken houses and other structures) as compared to the adult stages of their natural enemies (surrounding shrubs and other vegetation) led to a selective pesticide action through the differential treatment of the resting sites (Anderson and Poorbaugh 1964, Anderson 1965).

Selective action can be obtained by a variety of techniques taking advantage of diurnal and seasonal behavior, pesticide formulation, dosage rate, etc. (Stern et al 1959). Restricting the placement of insecticides can often achieve marked selectivity. Tobacco hornworms can be controlled by restricting treatment to the upper one-third of the plant, since the immature leaves are preferred by the larvae; this avoids disruption of the insect complex on other parts of the plant (NAS 1969, Guthrie et al 1959). The same type of selective action on a pest-parasite complex can be obtained by treating only those areas where the pest-parasite ratio is unfavorable. On an area-wide basis, the balance in favor of the hymenopterous parasite, *Apanteles medicaginis,* as compared to its host caterpillar was shifted in the supervised control program in California even though many parasites were destroyed in the treated areas.

The use of granular formulations in the central whorl of maize plants has a similar relationship with lepidopterous pests of maize and pests and their enemies on other parts of the plant (Ortega, unpublished data). It has long been known that the type of carrier and residue deposit may produce differential effects on the insect complex associated with a crop (Flanders 1941, Holloway and Young 1943).

The proper timing of insecticides is one of the best ways to produce a selective action on the pest and natural enemy complex. This often requires improved population monitoring techniques such as those which are now being provided by black-light and pheromone traps. Frequently, careful research on dosage rates and formulations are required to get the desired selectivity. Many currently recommended dosage rates for pesticides cause over-kills and non-selective results. These heavy treatments cover-up for a lack of precise information on timing, proper dosage rates and desirable level of control. In any case, for better selectivity an intimate knowledge of the behavior patterns of the pests and their natural enemies is required.

Another point to keep in mind in seeking this specificity is that as we increase dosages to high levels, there is a tendency to lose specificity. In the extreme case, at very high dosages there is literally no specificity. Under integrated control systems, often the population dynamics of the situation or the pest abundance-crop damage relationship is such that we do not need to attain a high degree of mortality. Instead of seeking 95% mortality or higher, we may find that a kill of 75% or even much lower results in adequate and sometimes longer lasting overall control. Under such circumstances, the dosage of pesticide needed for the low percent mortality permits the desired selective action.

Current pesticide screening procedures are so dependent on percent kill of a few test insects that many compounds of great potential value for crop protection have been passed by. We need a wide array of pesticides for crop protection, and we are dependent on the chemical industry for these materials. It is recognized that many materials of great importance for crop protection will have only limited use and hence may not be profitable to produce from the view of the chemical industry. The ecological aspects of pesticide use (and this is now a prime consideration of an aroused public) must be seriously evaluated and provided for in the future development of pesticides. Means must be sought (perhaps through federal agencies) to make selective and other potentially useful materials profitable to produce.

The ideal selective material is not one that eliminates all individuals of the pest species while leaving all of the natural enemies. The maintenance of low populations of pest species is, of course, essential to the continuity of predators or parasites. In Washington apple orchards, the apple rust mite, *Aculus schlechtendali,* can be tolerated at moderate populations without economic loss. These populations of apple rust mite are a good food source for mite predators which not only prevent eruptions of apple rust mite but also regulate McDaniel mite at low densities and help to control European red mite (Hoyt 1970).

After a shift to pesticide chemicals which are more selective in their action, it may be sometime before balance is restored. The effects of the previous treatments may last several years. In some instances, effective biological control no longer exists and would have to be re-established. This may be a slow process (DeBach 1951, Stern et al 1959). It took four years to re-establish normal predator-prey relationships for mites in apple orchards in the state of Washington after mite control sprays were modified (Hoyt 1970).

HUMAN HEALTH CONSIDERATIONS

It is obvious that we should not knowingly use pesticides in ways that would constitute a risk to human health. Such hazards can occur to the individual applying the pesticide, or to persons entering the treated area either during the application or at some appreciable time later, or to persons exposed to the residues of the pesticide on or in the harvested crops, or in other more subtle ways. When such a hazard is discovered, procedures should be taken to avoid the risk, e.g., proper masks and clothing for applicators, minimum time period after treatment before treated area can be entered, or minimum time after treatment before harvest. In those instances where the risk cannot be reduced to a safe level, then that particular pesticide use should be eliminated and a suitable alternative control measure be sought. I do not suggest that all uses of the pesticide be banned but rather that the particular use involved be eliminated. Furthermore, I stress again when alternatives are considered that all aspects, both positive and negative, be carefully weighted.

SIMPLIFICATION OF THE ECOSYSTEM

Most agro-ecosystems, if they have not been subjected to heavy use of pesticides, have a rather large complex of parasites, predators and pathogens that feed upon and are dependent upon pest species as well as upon an even larger number of potential pests. In addition, there are numerous other forms that do not stand in any immediate and direct relationship to the pest species. These, however, may be very important in providing alternative food sources for parasites and predators and thus in stabilizing and maintaining their numbers when the pest species population is at a low level, absent, or otherwise unavailable. Finally, there are other significant but so-called indifferent species (that is, indifferent to the abundance of the pest species). These include the scavengers, decomposers, pollinators, and other forms that may provide some useful function in the agro-ecosystem—although we do not always know their precise functions nor can we say that they must be present if the farmer is to obtain a profitable return.

In general, we should seek to preserve these rich, diverse biotic complexes, but we should also seek to understand the functions of the various elements. The fact that crop monocultures are often severely damaged by pests whereas the diverse climax vegetation of many natural environments is little harmed has led to the assumption by some that maximum diversity is desirable in agricultural areas. It is thought that this will preserve the stability inherent in the natural environment. There is an element of truth here but it may have been overstressed. Moreover, hedgerows and other "semi-natural" vegetation that adjoins crops are well known to be overwintering sites, sheltering places, and food sources of many crop pests. Such vegetation may also benefit natural enemies if it supports their alternative hosts, but this will not automatically provide a marked increase in natural biological control. Much depends on whether this counteracts benefits to the pest population. Once the delicate stability of the climax vegetation has been disturbed by man, if only slightly, the vegetation, although still complex, may nevertheless have been sufficiently altered to benefit a pest relative to the natural enemies that previously regulated it. (FAO 1968).

In the present state of knowledge it is difficult to generalize as each case requires thorough study. Thus, sometimes almost any form of diversity of plant species and age structure decreases pest damage (e.g., the spruce budworm, *Choristoneura fumiferana*), but other examples show that damage decreases with increasing simplification (e.g., where key secondary host plants are removed as for the Mediterranean fruit fly in Israel and Tunisia).

Diversity within the crop environment should be considered separately from diversity in the "semi-natural" environment adjacent to it. Within the cropped area, diversity can often be readily established and manipulated in terms of numbers of plant and animal species, plant age, and cultural practices. This is already important in integrated control, e.g., use of strip cutting of alfalfa; and strip cropping of cotton and other crops (van den Bosch and Stern 1969, Stern 1969) and will no doubt be increasingly important in future control practices. Nevertheless, the right kind of diversity must be established. Furthermore, the same kind of diversity can be harmful in one place and beneficial in another, e.g., in Tanzania, California, and elsewhere growing maize with cotton increases *Heliothis* damage to cotton, whereas in Perú the same kind of diversity helps to control *Heliothis*. In some circumstances the maize is the source and cause of damaging *Heliothis* attacks, but in others the conditions favor continuity of natural enemies and the establishment of a stable equilibrium between the pest and its enemies which keeps the pest scarce.

Each agro-ecosystem must necessarily be considered as a distinct man-manipulated system in which integrated control may involve the elimination of undesirable elements of diversity as well as maintaining or establishing desirable elements. The use of chemical pesticides is one of our most powerful tools in manipulating this diversity.

The introduction of chemicals into an agro-ecosystem may have subtle effects far more profound than the simple differential kill discussed previously. For example, when a mite egg has been treated with malathion the probing of the egg by a typhlodromid predator is not normal (Hussey and Bravenboer 1970). Neither host nor parasite have been killed by the pesticide, but the host-parasite relationship has been greatly altered.

The pesticides applied against a target insect may also have a striking effect on the crop plant which in turn affect the insect's or the plant's ability to produce. For example, repeated applications of certain organophosphates to the foliage or soil in early season cause cotton plants to develop vegetatively and result in delayed maturity of the bolls (van den Bosch et al 1970, Leigh 1963). These effects of pesticides may modify the crop plant as a suitable host for the pest insect. This often increases the fecundity and longevity of the pests involved (Chaboussou 1965). The implication for the population dynamics of pest species and for pest management are, of course, great.

PEST MANAGEMENT DECISIONS

Today many decisions with respect to pest management are being made in a political context and with little consideration of the fundamental technological facts upon which a sound pest management decision should be based. Moreover, it is nearly impossible at present to appraise the sociological effects and problems that lie at the root of the politics (Smith 1969a). Each pesticide usage should be judged on the basis of the potential positive values to be achieved by such usage as weighted against the possible negative values occurring from residues on the harvested crop, occupational hazards to humans, hazards to pollinating and other beneficial insects, effects on wildlife, the contribution to total environment pollution, etc. (Headley and Kneese 1969). Each use of a chemical must be judged independently. Banning all use of a chemical is unwise unless it is clear that all uses of that chemical are, in balance, harmful. Likewise it is irresponsible to advocate the total replacement of pesticides with sophisticated, but poorly tested alternative pest-management techniques. It is a disservice to society to discard the good crop protection methodology currently available and to adopt in its place a glamorous new but untested methodology. In the future development of pest management for a world agriculture it will be just as important to apply the methodology of traditional pest control techniques as it will be to find new revolutionary approaches, some of which may be expected to have a limited or no practical value (Smith 1969a).

ACKNOWLEDGMENTS

The manuscript for this paper was reviewed by my colleagues W. W. Allen, C. B. Huffaker, C. S. Koehler, P. S. Messenger, J. E. Swift, and R. van den Bosch. They made many helpful suggestions for which I am grateful. However, the full responsibility for the final statement should, of course, be placed on me.

REFERENCES CITED

Anderson, J. R. 1965. A preliminary study of integrated fly control on northern California poultry ranches. Proc. Calif. Mosq. Control Assoc. 33:42-44.

Anderson, J. R. and J. H. Poorbaugh. 1964. Observations on the ethology and ecology of various Diptera associated with northern California poultry ranches. J. Med. Ent. 1(2):131-47.

Brown, A. W. A. 1968. Insecticide resistance comes of age. Bull. Ent. Soc. Amer. 14:3-9.

Brown, Lester. 1970. Seeds of Change.

Chaboussou, F. 1965. Noureaux aspects de la phytiatrie et de la phytopharmacie. Le phénomè de la trophobiose. Proc. FAO Symposium on Integrated Pest Control 1:33-61.

De Bach, P. 1951. The necessity for an ecological approach to pest control on citrus in California. Jour. Econ. Ent. 44(4):443-47.

Eidt, D. C. and J. O. Nichols. 1970. An outbreak of a sawfly, *Pamphilius phyllisae,* on northern red oak with notes on larval morphology. Can. Ent. 102(1):53-63.

FAO. 1968. Report of the Second Session of the FAO Panel of Experts on Integrated Pest Control. PL/1968/M/3. Rome, 19-24 Sept., 1968.

FAO. 1969. Fifth Session of the FAO Working Party of Experts on Re-

sistance of Pests to Pesticides. Rome, Italy, September, 1968.
Flanders, S. E. 1941. Dust as an inhibiting factor in the reproduction of insects. Jour. Econ. Ent. 34(3):470-1.
Guthrie, F. E., R. L. Rabb, T. G. Bowery, F. R. Lawson, and R. L. Baron. 1959. Control of hornworms and budworms on tobacco with reduced insecticide dosage. Tob. Sci. 3:65-68.
Headley, J. C. and A. V. Kneese. 1969. Economic implications of pesticide use. Annals New York Acad. Sci. 160(1):30-39.
Holloway, J. K. and T. Roy Young, Jr. 1943. The influence of fungicidal sprays on entomogenous fungi and on the purple scale in Florida. Jour. Econ. Ent. 36(3):453-57.
Hoyt, S. C. 1970. The developing program of integrated control of pests of apple in Washington. A.A.A.S. Symposium on Theory and Practice of Biological Control. Boston, Dec. 30-31. 1969.
Hussey, N. W. and L. Bravenboer. 1970. Control of pests in glasshouse cultures by introduction of natural enemies. A.A.A.S. Symposium on Theory and Practice of Biological Control. Boston, Dec. 30-31, 1969.
Leigh, T. F. 1963. The influence of two systemic organophosphates on growth, fruiting and yield of cotton in California. Jour. Econ. Ent. 56(4):517-22.
National Academy of Sciences. 1969. Insect-pest Management and Control. Principles of plant and animal pest control. Vol. 3, 508 pp.
Smith, Ray F. 1969a. The new and the old in pest control. Accad. Naz. dei Lincei 366(128):21-30.
Smith, Ray F. 1969b. The importance of economic injury levels in the development of integrated pest control programs. Qual. Plant. Mater. Veg. 17(2):81-92.
Smith, Ray F. and R. van de Bosch. 1967. Integrated Control. *In* Kilgore, W. W. and Doutt, R. L. (ed.). Pest Control. Biological, Physical, and Selected Chemical Methods. Academic Press, N. Y. pp. 295-340.
Stern, V. M. 1969. Interplanting alfalfa in cotton to control lygus bugs and other insect pests. Proc. Tall Timbers Conf. on Ecological Animal Control by Habitat Management 1:55-69.
Stern, V. M., Ray F. Smith, R. van den Bosch and K. S. Hagen. 1959. The integrated control concept. Hilgardia 29(2):81-101.
Swift, J. E. 1970. Unexpected effects from substitute pest control methods. Symposium on the Biological Impact of Pesticides in the Environment, Oregon State Univ., Corvallis, Aug. 18-20, 1969.
van den Bosch, R., T. F. Leigh, V. M. Stern, L. A. Falcon and D. Gonzales. 1970. The developing program of integrated control of cotton pests in California. A.A.A.S. Symposium on Theory and Practice of Biological Control. Boston, Dec. 30-31, 1969.
van den Bosch, R. and V. M. Stern 1969. The affect of harvesting practices on insect populations in alfalfa. Proc. Tall Timbers Conf. on Ecological Animal Control by Habitat Management 1:47-54.
Waterhouse, D. F. 1969. The importance of the integrated control concept in dealing with the problems of pesticide resistance. FAO Symposium on Pest Resistance. Rome, Italy, September, 1969.

SAMPLING METHODS

Sequential Sampling in Forest Insect Surveys

BY

W. E. WATERS

SEQUENTIAL ANALYSIS was first developed in 1943 for use in rapid evaluation of war research problems (Wald, 1943). First classified as Restricted, it was introduced into a large number of war plants as a basis for acceptance inspection by the Army, the Navy, and the Office of Scientific Research and Development. Its practical application to the fields of both sampling inspection and experimentation was quickly evident, and it was cleared for wider use in 1945. Wald (1945) and the Columbia University Statistical Research Group (1946) developed general formulas for the types of problems most commonly met.

The application of sequential analysis to biological problems has been somewhat limited,[1] but its possibilities are becoming evident—particularly in forestry and related fields. Sequential sampling serves to classify populations rather than to provide estimates of population parameters; so it is particularly applicable to surveys. It involves a flexible sample size in contrast to conventional sampling procedures, which usually specify a fixed number of sampling units to be examined. Units chosen at random from a sample are examined in sequence until the sample falls into one or more classes distinguished by specified limits. The items in or on each sampling unit (insects, in the case of forest-insect surveys) are either counted or simply recorded as present or absent, and just enough time is spent with each lot to label it properly.

Very logically, but rather oddly at first glance, the least work is required where population levels are either very low or very high; more sampling is required at intermediate levels. Adequate fixed-sample-size methods applied to surveys invariably result in too little time and labor spent at sampling points or lots having low populations, and too much time where populations are high. Usually the savings obtained by the sequential method are considerable, often exceeding 50 percent (Statistical Research Group, 1946 — hereafter called SRG). Even where practical limitations make it necessary to set an upper limit on sample size, sequential sampling can be used effectively and will result in substantial savings (Bliss and Calhoun, 1954).

Application to Forest Insect Surveys

Surveys are the first line of attack in minimizing losses by destructive forest insects. They enable forest entomologists to determine the current status of known pests, where and when outbreaks are going to oc-

[1]Oakland, 1950; Morgan, *et al.*, 1951; Bross, 1952; Stark, 1952; Morris, 1954.

FOREST SCIENCE, 1955, vol. 1, no. 1, pp. 68-79.

The author is Entomologist stationed at the Forest Insect and Disease Laboratory, New Haven, Conn., of the Northeastern Forest Experiment Station, Forest Service, U. S. Dept. of Agriculture.

cur, if and where direct control measures are needed, how effective these measures have been, and where utilization cuttings are needed to prevent excessive losses (Orr, 1954).

The Forest Pest Control Act, passed by Congress in 1947, has made possible more active cooperation between the Federal Government and the states, private industry, and individual timberland owners. At the same time, however, it has placed a greater responsibility and work load on the Federal forest-insect-survey organization, already limited as to funds and manpower, for conducting surveys and lending technical advice and assistance in control projects. Improved efficiency of survey techniques is one means of better meeting this obligation, and sequential sampling can contribute much toward that end.

The sequential method is most feasible for reconnaissance surveys, whose function is to determine the extent and severity of known infestations and the need for direct control measures. For this type of survey both speed and reliability are essential. The infestation, or portions of it, is usually put into classes such as light, medium, or heavy —which purpose sequential sampling will serve most efficiently.

Also, for determining whether or not control measures are necessary, considerable time and expense can be saved if arbitrary limits are set for just two classes: "needing control" and "not needing control." These class limits, of course, must be based on a fundamental knowledge of the insect and the forest conditions present. The limits may be in terms of number of insects per sampling unit. For example, it might be judged necessary to spray against a certain defoliator when populations reach a level of 20 or more insects per branch sample, while 10 or less insects per branch would not be sufficient cause for such action.

Or the limits may be in terms of percentage of trees or area infested. It might be stated, for example, that white pine plantations having 15 percent or more trees currently damaged by the white pine weevil should be treated, while those having 10 percent weeviling or less should not. The active destructive stage of the insect may be sampled as a basis for judging the seriousness of the situation, or egg surveys may be conducted to obtain estimates of the potential degree of attack.

Similarly, the sequential method can be used for post-control appraisal purposes to quickly determine whether or not satisfactory control has been obtained and, if not, what trees or areas require re-treatment or continued surveillance. Also, as an adjunct to aerial surveys, which though much faster and more economical than ground methods still require supplemental ground information for accurate interpretation, sequential sampling will permit better and more rapid coverage of an area.

Other applications will arise as situations warrant. It is often important to know whether or not a natural control factor known to be present — a certain parasite, say — is at a level sufficient to alter the trend of an infestation. If enough is known of the parasite to specify that level, the decision will be reached most quickly and economically by sequential sampling.

It should be strongly emphasized that the sequential method, like any method of sampling, is only a tool and is, moreover, subject to error. It will make the best use of the competent observer's knowledge of an insect, the forest, and the basic dynamic relationships between them; but every sequential plan must be properly developed and applied to provide a sound basis for decisions.

Sequential Plans: Values and Errors Involved

Certain fundamental information is necessary to develop a sequential plan for sampling an insect population. First, the mathematical distribution of the insect counts (or other quantities) on the sampling unit used must be known. In certain cases this may be assumed with little chance of being wrong. Forest-insect-survey data will gen-

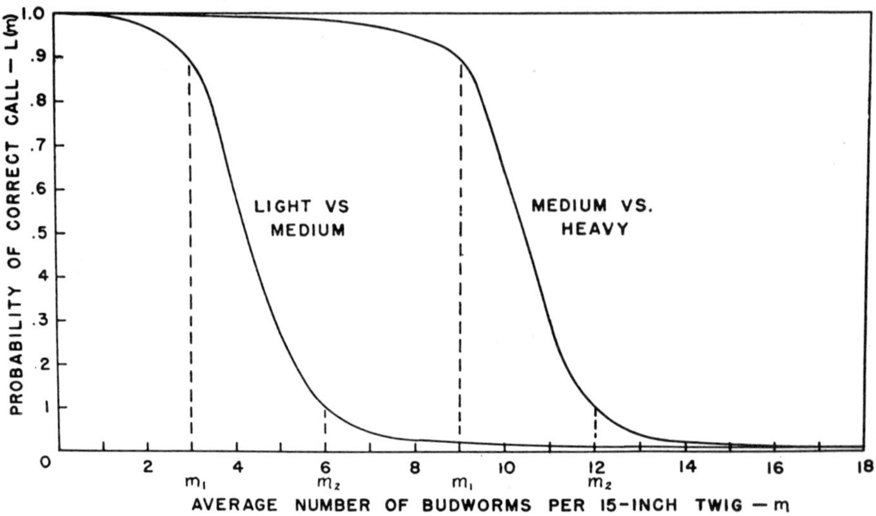

Fig. 1. Operating Characteristic curves of the sequential plan for sampling spruce budworm larval populations (4th to 6th instars) on 15-inch balsam fir twigs—90 percent confidence level. The plotting points of L(m) and m for these curves were obtained by inserting values of a "dummy" variable x in the following formulas (Oakland, 1950):

$$L(m) = \frac{A^x - 1}{A^x - B^x} \text{ where } A = \frac{1-\beta}{\alpha} \text{ and } B = \frac{\beta}{1-\alpha}$$

$$p = \frac{1 - \left(\dfrac{q^1}{q^2}\right)^x}{\left(\dfrac{p^2q^1}{p^1q^2}\right)^x - 1} \quad \text{and } m = kp$$

erally fit one of the four following distributions: binomial, negative binomial, Poisson, or normal. Procedures for determining which distribution a given set of data will best fit are fully described in numerous statistical texts, such as those by Snedecor (1946) and Fisher (1948).

Next, enough must be known of the insect and the damage that it does to set class limits that have a practical meaning. The two or more classes (or hypotheses) that are to be distinguished, and their differences, must be realistic. Most commonly these are based on the causal relationship between the number of insects per sampling unit and the average amount of damage done by them. However, the tolerance limits for decisions such to spray or not to spray are based more on a judgment of maximum allowable damage and are logically influenced by economics as well as biological considerations.

Two kinds of errors are involved in sequential sampling (they are present also in conventional methods applied to similar situations, though not always recognized). There is the risk α of rejecting a hypothesis when it is true, for example, calling a light infestation medium; and the risk β of accepting it when the alternative is true,

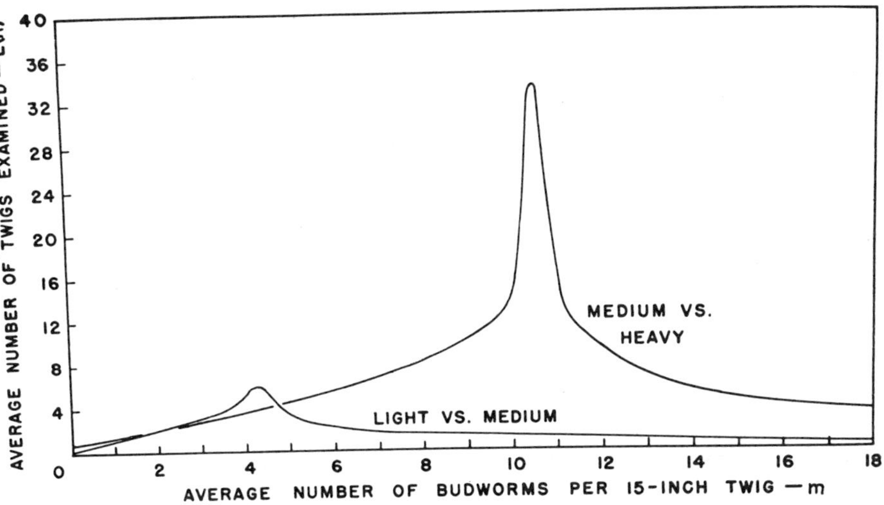

The computed values of L(m) and m are given in Table 2.
Figure 2. The Average Sample Number curves of the sequential plan for sampling spruce budworm larval populations (4th to 6th instars) on 15-inch balsam fir twigs—90 percent confidence level. The values defining these curves were computed by the following formulas (8):

$$L(m) = \frac{A^x - 1}{A^x - B^x} \text{ where } A = \frac{1 - \beta}{\alpha} \text{ and } B = \frac{\beta}{1 - \alpha}$$

$$E(n) = \frac{h_2 + (h_1 - h_2)\, L(m)}{m - b}$$

The computed values of L(m) and E(n) are listed in Table 2.

that is, calling a medium infestation light. The confidence levels expressed by the values assigned to these risks must be set with a practical eye to the feasibility of obtaining the specified degree of reliability and the use to which the information will be put.

Two other features of a sequential plan, the Operating Characteristic curve (OC curve) and the Average Sample Number curve (ASN curve), may be described here also. For each plan it is sometimes important to know in advance what the chances are of correctly labeling a sample or of making the correct decision as to spray or not to spray, and so forth, at the various population or infestation levels that are expected. For example, if a plan specifies that two insects or less per sampling unit constitute a light infestation and five insects or more a medium infestation, and that α and β are 0.10, then the probability of correctly calling an infestation light when there is an average of two insects per unit is 0.90 and the probability of calling it light when there are five insects per unit is 0.10. These and other probabilities of correct calls for different population levels (mean counts) or infestation percentages can be easily computed and shown graphically as an OC curve (Fig. 1).

The sampling unit and the work involved in obtaining it should be familiar to the observer. For most survey purposes the odds of 9 to 1 (α and $\beta = 0.10$) are satisfactory. On the average, then, 1 out of every

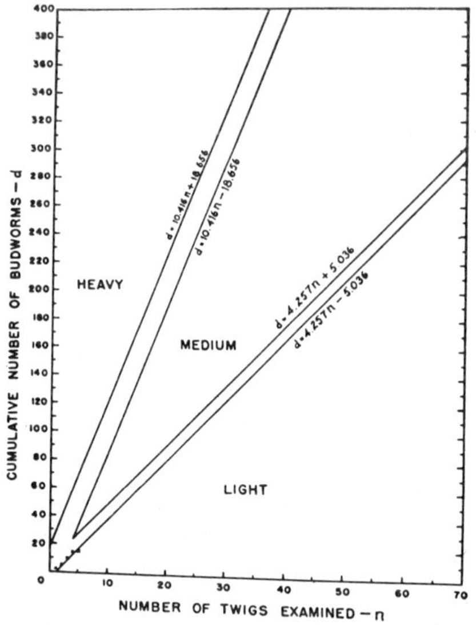

Fig. 3. Sequential graph for sampling spruce budworm larval populations (4th to 6th instars) on 15-inch balsam fir twigs—90 percent confidence level.
The small x's represent successive counts of 2, 3, 5, 4, and 1 budworm larvae. As the fifth count puts the cumulative total (15) below the decision line for LIGHT, sampling ceases and the sample is labeled LIGHT.

10 sampling points or lots may be mislabeled. The risk values for α and β need not always be equal. With some insects or under certain conditions it might be considered a more serious error to call a heavy infestation medium than to identify a medium infestation as heavy, in which case β would be set at 0.05 while α remained at 0.10. Then the chances of mislabeling a heavy infestation would be just 1 in 20. For post-control appraisals both risks should probably be set at 0.05 or lower.

Also, although the sequential method involves no fixed sample size, it is possible to calculate for each sequential plan the *average* sample size to be expected at different population levels. This can be shown graphically as the ASN curve (Fig. 2). The average sample size is directly related to the allowance or gap between classes and the desired degree of reliability. Narrowing the gap and lowering the risks of error will increase the average amount of sampling required. These factors should be weighed carefully in figuring the cost of any survey.

The curve formulas for the negative binomial distribution and the computational procedure are given in the example concerning spruce budworm populations. To avoid confusion with the general sequential formulas, however, the OC and ASN curve formulas for the other distributions are given in the Appendix.

General Formulas

Every sequential plan is defined by one or more pairs of parallel lines (Fig. 3). These lines, or values taken from them, are the criteria for distguishing between classes of infestation or for making decisions as to necessary action. A plan can be presented graphically, or for more convenient use in the field, set up in tabular form (Table 1).

The general formula for any pair of parallel decision lines is:

$$d_{1,2} = bn \pm h_{1,2}$$

where d_1 = the maximum value (s) for the lower class in terms of *cumulative* number of insects, units with insects present, or damaged trees; d_2 = the minimum value (s) for the higher class in terms of *cumulative* number of insects, etc.; n = number of sampling units examined; h_1 *and* h_2 = intercepts (where the respective lines cross the vertical axis of the graph); and b = slope of the lines.

For solving the line equations two other values, mentioned previously, are always involved but are entered in various ways depending on the distribution concerned: (1) the assigned risks, α and β, and (2) the class limits, designated hereafter as m_1 and m_2. In solving these equations m_2 must always be larger than m_1.

Also, in the computations for the following distributions the slope b is always between m_1 and m_2, and the intercepts h_1 and h_2 are different in sign but are equal when $\alpha = \beta$.

Binomial distribution. This is the distribution to assume when each observation is recorded merely as insects present or absent (regardless of number) or damage present or absent, and the result is acceptance or rejection of a hypothesis. Every sampling unit in a lot should have the same chance of insects occurring on it. A plan would be based on this distribution, for example, if it were to be used to determine whether or not an infested area needed control. The limits for distinguishing "needing control" from "not needing control" would be stated as hypotheses in percentage terms—20 percent or less of the trees infested, no spray; 30 percent or more infested, spray. The application of sequential sampling to a problem of insect parasitism, based on the binomial case, has been discussed by Oakland (1951).

Table 1. Sequential table for sampling spruce budworm larval populations (4th-6th instars) on 15-inch balsam fir twigs[1]

(90 percent confidence level)

Number of twigs examined	Cumulative number of budworms					
	Light		Medium	Medium		Heavy
(1)	(2)		(3)	(4)		(5)
1	----		----	----		29
2	3		----	----		39
3	8		----	----		50
4	12		22	23		60
5	16		26	33		71
6	20		31	44		81
7	25		35	54		92
8	29		39	65		102
9	33		43	75		112
10	38		48	85		123
11	42		52	96		133
12	46	CONTINUE SAMPLING	56	106	CONTINUE SAMPLING	144
13	50		60	117		154
14	55		65	127		164
15	59		69	138		175
16	63		73	148		185
17	67		77	158		196
18	72		82	169		206
19	76		86	179		217
20	80		90	190		227
21	84		94	200		237
22	89		99	210		248
23	93		103	221		258
24	97		107	231		269
25	101		111	242		279

[1]This table is used as follows: Sampling is continued only so long as the cumulative total count of budworms is between the listed values for LIGHT and MEDIUM (columns 2 and 3) or between those of MEDIUM and HEAVY (columns 4 and 5), or where a dash is shown. For example, if counts of 2, 3, 5, 4, and 1 budworms on successive twigs occurred, sampling would cease after the last (the 5th) twig because on that twig the count (15) fell below the limit value for LIGHT, and the sample would be so labelled. Similarly, when distinguishing between a MEDIUM and HEAVY population (columns 4 and 5), sampling is continued until the cumulative count of budworms for any given number of twigs is equal to or less than the count listed MEDIUM (column 4) or equal to or higher than that listed under HEAVY (column 5). A MEDIUM population is determined when the cumulative budworm count for any number of twigs (that is, 4 or more) is equal to or higher than the count listed for MEDIUM (column 3) and equal to or lower than that listed for MEDIUM (column 4). As indicated, it is possible to identify a LIGHT population after examining just 2 twigs, a MEDIUM population after 4 twigs, while a count of 29 or more budworms on the first twig would immediately specify a HEAVY population.

The equations for the decision lines are:

$$d_1 = bn - h_1 \text{ (lower line)}$$
$$d_2 = bn + h_2 \text{ (upper line)}$$

where d stands for the number of sampling units with insects present or trees with damage evident; n, the number of units examined; h_1 and h_2, the intercepts; and b, the slope of the lines.

The slope and intercepts are calculated by the following formulas (SRG, 1946):

$$b = \frac{\log\left(\frac{1-m_1}{1-m_2}\right)}{\log \frac{m_2}{m_1}\left(\frac{1-m_1}{1-m_2}\right)}$$

$$h_1 = \frac{\log\left(\frac{1-\alpha}{\beta}\right)}{\log \frac{m_2}{m_1}\left(\frac{1-m_1}{1-m_2}\right)}$$

$$h_2 = \frac{\log\left(\frac{1-\beta}{\alpha}\right)}{\log \frac{m_2}{m_1}\left(\frac{1-m_1}{1-m_2}\right)}$$

Negative binomial distribution. Bliss and Fisher (1953) found this distribution applicable to a wide range of biological data involving counts of plants or animals in units of space or time. Most forest-insect-survey data in the form of counts per sampling unit will follow this distribution. It is a type of "contagious" distribution, where the presence of one insect on a unit increases the chance of others occurring on that unit. This commonly results from larvae hatching from eggs laid in masses, or from gregarious feeding habits, although other factors may be responsible also.

In a negative binomial series the variance varies with the mean, but disproportionately so. This characteristic of the variance being significantly larger than the mean has been called "over dispersion" (Bliss and Fisher, 1953). The distribution is described by two parameters, the mean m and an exponent k. As the variance approaches the mean (and k approaches infinity) or as the population level decreases, the negative binomial will tend toward the Poisson distribution. Populations at low density may, in fact, fit the Poisson series more closely.

The k is theoretically independent of the mean, so that if a sequential plan is to be applied to a number of localities or over a wide area comprising different population levels, a pooled estimate of k should be obtained. Methods of estimating a common k for different sets of counts of an insect have been described by Anscombe (1949) and very recently by Bliss and Owen.[2]

The occurrence of both larvae and egg masses of the spruce budworm on 15-inch-twig sampling units from balsam fir trees agrees well with the negative binomial, and the sequential plan for sampling budworm larval populations that it described later in this paper was devised on that basis. Morris (1954) has developed sequential plans, based on the negative binomial case, for sampling spruce budworm egg-mass populations in New Brunswick on whole-branch sampling units. Oakland (1950) has described an application of sequential sampling to the inspection of whitefish for tapeworm cysts, the distribution of which also follows the negative binomial.

The decision line equations for this series are:

$$d_1 = bn + h_1 \text{ (lower line)}$$
$$d_2 = bn + h_2 \text{ (upper line)}$$

where d is the cumulative number of insects, n is the number of sampling units examined, b is the slope of the lines, and h_1 and h_2 are intercepts. The slope and in-

[2]Bliss, C. I., and A. R. G. Owen, 1954. Estimating k for a series of negative binomial distributions. (Unpub. manuscript; paper presented at meeting of Biometric Society, Gainesville, Fla.)

tercepts are calculated as follows (Oakland, 1950):

$$b = k \frac{\log \frac{q_2}{q_1}}{\log \frac{p_2 q_1}{p_1 q_2}}$$

$$h_1 = \frac{\log \frac{\beta}{1-\alpha}}{\log \frac{p_2 q_1}{p_1 q_2}}$$

$$h_2 = \frac{\log \frac{1-\beta}{\alpha}}{\log \frac{p_2 q_1}{p_1 q_2}}$$

$$\text{where } p_1 = \frac{m_1}{k} \text{ and } q_1 = 1 + p_1$$

$$p_2 = \frac{m_2}{k} \text{ and } q_2 = 1 + p_2$$

Poisson distribution. Insect counts at low population densities will often follow this distribution. The counts in this series, unlike those in the negative binomial and other contagious distributions, are distributed independently and at random among the sampling units. There should be no tendency to group or cluster on the same units, nor should the presence of one insect prevent the occurrence of others on that unit. Stated differently, the units in every sample lot must be equally subject to attack, that is, must have the same chance of containing insects. The Poisson is described by just one parameter, the mean. The variance is equal to the mean or close to it at all population levels.

The equations for the decision lines are:

$$d_1 = bn - h_1 \text{ (lower line)}$$
$$d_2 = bn + h_2 \text{ (upper line)}$$

where d is the cumulative number of insects, n is the number of sampling units examined, b is the slope, and h_1 and h_2 are the intercepts. The slope and intercepts are given by the following formulas[3] (SRG):

$$b = \frac{m_2 - m_1}{\log_e m_2 - \log_e m_1}$$

$$h_1 = \frac{\log_e \frac{1-\alpha}{\beta}}{\log_e m_2 - \log_e m_1}$$

$$h_2 = \frac{\log_e \frac{1-\beta}{\alpha}}{\log_e m_2 - \log_e m_1}$$

Normal distribution. This distribution, common or generally assumed in most experimental work, is probably the least frequent, however, in forest-insect sampling. With normally distributed populations the presence or absence of an insect on a sampling unit is in no way related to the presence or absence of others. The variance and mean, moreover, are also independent. Stark (1952) developed a sequential plan for sampling populations of the lodgepole needle miner based on a normal distribution of the insect on the sampling units used.

The decision line equations for this distribution are:

$$d_1 = bn - h_1 \text{ (lower line)}$$
$$d_2 = bn + h_2 \text{ (upper line)}$$

where d is the cumulative number of insects, n is the number of units examined, b is the slope, and h_1 and h_2 are the intercepts. The formulas for the slope and intercepts (Stark, 1952) are:

$$b = \frac{m_1 + m_2}{2}$$

$$h_1 = \frac{s^2 \left(\log_e \frac{1-\alpha}{\beta} \right)}{m_2 - m_1}$$

$$h_2 = \frac{s^2 \left(\log_e \frac{1-\beta}{\alpha} \right)}{m_2 - m_1}$$

[3]Logarithms denoted $\log_e$ are natural logarithms to the base 2.71828 and are 2.30259 times the corresponding common logarithms to the base 10.

where s^2 is the estimated variance of the population. This can be obtained from the preliminary samples taken to determine the distribution of the data.

Application to Spruce Budworm

The standard survey procedure for sampling larvae and egg-mass populations of the spruce budworm in the Northeastern States has been to examine five 15-inch twigs from each of five balsam fir trees at every collection point. As with probability sampling generally, the sampling points (collection points) and the sample trees at these points must be selected without personal bias, preferably by random selection. The 25 twigs are cut by pole-pruner from branch terminals at approximately mid-crown height. As the pole-pruners used by the survey crews can be extended only about 25 feet, samples taken from large trees are generally from the lower portion of the crown. On small trees the twig samples tend to be from the upper crown. The extremes in tree size are avoided, however, so for practical purposes the mid-crown level is considered the sampling universe for the survey collections.

In 1953 sequential plans were devised for sampling both larvae (4th to 6th instars) and egg masses, and each plan was tested for efficiency and reliability against the standard method.[4] With an upper limit of 25 twigs, the sequential method showed a 40-percent saving in twig-examination time for both larvae and egg masses, with no loss of reliability in correctly identifying the degree of infestation for each. No discrepancies between the two methods occurred over the 50 points sampled for larvae. In the egg-mass sampling just 2 out of 71 points were mislabeled, and both of these were between medium and heavy and could have been called either way.

[4]Waters, W. E. 1953. A study of the application of sequential sampling to spruce budworm populations. U.S. Dept. Agr. Forest Insect Lab. Ann. Rpt. 1953: p. 14. 18 pp. (Unpub.)

Based on new data obtained in 1953, revised sequential plans have been developed for the larvae and egg-mass surveys. The plan for distinguishing light, medium, and heavy populations of spruce budworm larvae (4th to 6th instars) is outlined below.

Light and Medium Populations

Class limits:

Light—3 larvae or less per 15-inch twig.
Medium—6 larvae or more per 15-inch twig.

Confidence levels:

1. Risk of calling a light infestation medium, $\alpha = 0.10$
2. Risk of calling a medium infestation light, $\beta = 0.10$

Pertinent values:

$m_1 = 3$	$m_2 = 6$
$p_1 = 0.415$	$p_2 = 0.830$
$q_1 = 1.415$	$q_2 = 1.830$
$k = 7.228$[5]	$k = 7.228$[5]

[5]Common k derived by new method of Bliss and Owen. Grateful acknowledgment is made to C. I. Bliss for the computation of this k, for suggestions concerning the negative binomial series, and for permission to describe briefly the procedure for obtaining a provisional k (suitable for survey sequential plans) by this method, as yet unpublished. The steps are as follows:

1. Compute the variance and mean of a set of representative samples (the 25-twig samples in this case).
2. Tabulate values of $s^2 - \bar{y}$ and $y^2 - \frac{s^2}{n}$, where s^2 is the variance, $\bar{y}$ the mean, and n the number of units in the sample.
3. Plot above values on graph paper with $s^2 - \bar{y}$ on the vertical axis and $\bar{y}^2 - \frac{s^2}{n}$ on the horizontal axis.
4. If a few values are very discrepant they may be omitted. Calculate the mean values of $s^2 - \bar{y}$ and $\bar{y}^2 - \frac{s^2}{n}$, designated $\bar{y}'$ and $\bar{x}'$, respectively.
5. Draw a line from the 0 origin through the intersect of $\bar{y}'$, $\bar{x}'$.
6. Comp. $k = \frac{1}{\text{slope}}$, where the slope $= \frac{\bar{y}'}{\bar{x}'}$.

$$\text{The slope, } b = k \frac{\log \frac{q_2}{q_1}}{\log \frac{p_2 q_1}{p_1 q_2}} = 7.228 \frac{\log \frac{1.830}{1.415}}{\log \frac{(.830)(1.415)}{(.415)(1.830)}} = 4.257$$

and the intercepts,

$$h_1 = \frac{\log \frac{\beta}{1-\alpha}}{\log \frac{p_2 q_1}{p_1 q_2}} = \frac{\log \frac{.10}{1-.10}}{\log \frac{(.830)(1.415)}{(.415)(1.830)}} = -5.036$$

$$\text{and } h_2 = \frac{\log \frac{1-\beta}{\alpha}}{\log \frac{p_2 q_1}{p_1 q_2}} = \frac{\log \frac{1-.10}{.10}}{\log \frac{(.830)(1.415)}{(.415)(1.830)}} = 5.036$$

Decision lines:

The population is light if $d_1 \leqq 4.257\,n - 5.036$[6]

The population is medium if $d_2 \geqq 4.257\,n + 5.036$[6]

Medium and Heavy Populations

Class limits:

Medium—9 larvae or less per 15-inch twig.

Heavy—12 larvae or more per 15-inch twig.

Confidence levels: α and $\beta = 0.10$

Pertinent values:

$m_1 = 9$ $m_2 = 12$

$p_1 = 1.245$ $p_2 = 1.660$

$q_1 = 2.245$ $q_2 = 2.660$

$k = 7.228$ $k = 7.228$

The slope, $b =$

$$7.228 \frac{\log \frac{2.660}{2.245}}{\log \frac{(1.660)(2.245)}{(1.245)(2.660)}} = 10.416$$

and the intercepts,

$$h_1 = \frac{\log \frac{.10}{1-.10}}{\log \frac{(1.660)(2.245)}{(1.245)(2.660)}} = -18.656$$

$$h_2 = \frac{\log \frac{1-.10}{.10}}{\log \frac{(1.660)(2.245)}{(1.245)(2.660)}} = 18.656$$

Decision lines:

The population is medium if $d_1 \leqq 10.416n - 18.656$

The population is heavy if $d_2 \geqq 10.416n + 18.656$

The revised sequential plan applied to the 1953 survey data would have effected a 70-percent saving in time spent examining twigs for larvae with just 2 out of the 50 collection points mislabeled. Both of these incorrect calls would have been in the range between medium and heavy and could reasonably have been called either way. The plan is shown in graphic form in Figure 3 and in tabular form in Table 1.

Summary

Sequential analysis, a technique first developed in 1943 to expedite sampling in war research studies, is proving applicable to a wide range of biological data, including forest insect surveys. General formulas for devising sequential plans are given for the four mathematical distributions most commonly fitting forest insect survey data: binomial, negative binomial, Poisson, and normal. An application of this new

[6]The symbol $\leqq$ reads "is equal to or less than"; the symbol $\geqq$ reads "is equal to or more than."

Table 2. Computed values for the Operating Characteristic (OC) and Average Sample Number (ASN) curves of the sequential plan for spruce budworm larvae

Dummy variable	Light vs. medium populations				Medium vs. heavy populations			
x	p	m	L(m)	E(n)	p	m	L(m)	E(n)
∞	0	0	1.00	1.18	0	0	1.00	1.79
1	0.42	3.00	.90	3.21	1.25	9.00	.90	10.54
1/2	.50	3.61	.75	3.89	1.33	9.63	.75	11.87
0	.59	4.26	.50	5.96	1.44	10.42	.50	33.41
—1/2	.71	5.10	.25	2.99	1.54	11.11	.25	13.44
—1	.83	6.00	.10	2.31	1.66	12.00	.10	9.42
—3/2	.98	7.08	.04	1.64	1.78	12.87	.04	7.01
—∞	∞	∞	0	0	∞	∞	0	0

technique for determining spruce budworm larval populations is described. The computational procedures are outlined and the sequential plan is presented in graphic and tabular form. Operating Characteristic curves and Average Sample Number curves for this plan are also shown.

Literature Cited

ANSCOMBE, F. J. 1949. The statistical analysis of insect counts based on the negative binomial distribution. Biometrics 5: 165-173.

BLISS, C. I., and D. W. CALHOUN. 1954. Outline of biometry. Yale Coop. Corp. New Haven. 330 pp. (Proc.)

BLISS, C. I., and R. A. FISHER. 1953. Fitting the negative binomial distribution to biological data and note on the efficient fitting of the negative binomial. Biometrics 9: 176-200.

BROSS, I. 1952. Sequential medical plans. Biometrics 8: 188-205.

FISHER, R. A. 1948. Statistical methods for research workers. 10th ed. (rev.). Oliver and Boyd, London. 354 pp.

MORGAN, MAX E., P. MACLEOD, E. O. ANDERSON, and C. I. BLISS. 1951. A sequential procedure for grading milk by microscope counts. Conn. (Storrs) Agric. Exp. Sta. Bull. 276. 35 pp.

MORRIS, R. F. 1954. A sequential sampling technique for spruce budworm egg surveys. Canad. J. Zool. 32: 302-313.

OAKLAND, G. B. 1950. An application of sequential analysis to whitefish sampling. Biometrics 6: 59-67.

————. 1951. Sequential analysis. Canada Dept. Agr., Biometrics Unit, Ottawa. 8 pp. (Proc.)

ORR. L. W. 1954. The role of surveys in forest insect control. J. For. 52: 250-252.

SNEDECOR, G. W. 1946. Statistical methods. 4th Ed. Iowa St. Coll. Press, Ames. 485 pp.

STARK, R. W. 1952. Sequential sampling of the lodgepole needle miner. Forestry Chron. 28: 57-60.

STATISTICAL RESEARCH GROUP, COLUMBIA UNIVERSITY. 1946. Sequential analysis of statistical data: applications. Columbia Univ. Statis. Res. Gp. Rpt. 255 (rev.) and Office Sci. Res. and Devlpmt. Applied Math. Panel Rpt. 30.2R (rev.). Columbia Univ. Press. 393 pp.

WALD, A. 1943. Sequential analysis of statistical data: theory. Columbia Univ. Stat. Res. Gp. Rpt. 75 and Office Sci. Res. and Devlpmt. Rpt. 1998.

WALD, A. 1945. Sequential tests of statistical hypotheses. Ann. Math. Statis. 16: 117-186.

Appendix

Formulas for Computing the Operating Characteristic Curve and the Average Sample Number Curve

Binomial Distribution

OC curve

$$(1)\ p' = \frac{x^{s}-1}{x-1},\ Lp' = \frac{x^{h_1+h_2} - x^{h_1}}{x^{h_1+h_2} - 1}$$

$$(2)\qquad p'' = p'\,x^{1-s},\ Lp'' = \frac{Lp'}{x^{h_1}}$$

When various values are assigned to the dummy variable x in the equations above, formula 1 gives that portion of the curve for which p lies between 0 and b and formula 2 gives that portion of the curve for which p lies between b and 1. For plotting, $L(m)$ will be the computed values of Lp' and Lp'', and m will be the successive values of Kp' and Kp''. The quantity x should always be greater than 1.

If α and β are the same, a single formula for p which does not involve a dummy variable can be used:

$$(3)\quad p = \frac{\left(\dfrac{Lp}{1-Lp}\right)^{\frac{b}{h}} - 1}{\left(\dfrac{Lp}{1-Lp}\right)h - 1}$$

where $h = h_1 = h_2$

ASN curve

$$E(n) = \frac{Lm\,(h_1 + h_2) - h_2}{b - p}$$

Poisson Distribution

OC curve

$$(1)\ m' = \frac{b \log_e x}{x-1},\ Lm' = \frac{x^{h_2} - 1}{x^{h_2} - x^{-h_1}}$$

$$(2)\quad m'' = m'x,\ LM'' = x^{-h_1}\,Lm'$$

Substitution of various values of x in the equations above gives successive points on the OC curve. In these equations also, x should always be greater than 1. The plotting values of $L(m)$ and m will be the computed values of Lm', Lm'' and m', m'', respectively.

When $\alpha = \beta$ (and $h_1 = h_2$) a single formula for m can be used:

$$(3)\qquad m = \frac{\dfrac{b}{h}\log_e \dfrac{L(m)}{1-L(m)}}{\left(\dfrac{L(m)}{1-L(m)}\right)^{\frac{1}{h}} - 1}$$

ASN curve

$$E(n) = \frac{L(m)\,(h_1 + h_2) - h_2}{b - m}$$

Normal Distribution

OC curve

$$(1)\quad K = \frac{2\,(b-m)}{S^2}$$

$$(2)\quad t_1 = h_1 K$$

$$(3)\quad t_2 = (h_1 + h_2)K$$

$$(4)\quad L(m) = \frac{e^{t_1} - 1}{e^{t_2} - 1}$$

The plotting values of $L(m)$, m are calculated by assigning various values of m in equation 1 and computing the corresponding $L(m)$'s from equation 4. The quantities e^{t_1} and e^{t_2} are computed by use of *Tables of Exponential Function* e^x, Federal Works Agency, Works Project Administration for the City of New York, National Bureau of Standards, 1939.

ASN curve

$$E(n) = \frac{L(m)\,(h_1 + h_2) - h_1}{m - b}$$

SAMPLING AS A BASIS FOR PEST MANAGEMENT STRATEGIES

D. GONZÁLEZ

THE ECOLOGICAL approach to habitat management theme of this conference is especially appropriate in a year which has seen such intense—if belated—public and scientific concern over the preservation of environmental quality. But far more than concern is demanded if we are to effect a qualitative improvement of our environment and contain the spread of environmental deterioration. Those of us who are in the biological sciences must be prepared to reexamine our past approaches to habitat management, realistically appraise both their strengths and weaknesses, and begin seeking new strategies that more fully recognize the complexity and interrelatedness of all biological components of an environment.

The need for new management strategies is particularly evident in the discipline of economic entomology, where traditional emphasis has been on eliminating target pest species, especially through chemical means. The achievements of such unilateral methods at times have been spectacularly effective in increasing our food and fiber supply, but only recently have we become aware of the correspondingly detrimental and cumulative consequences in terms of increasing costs, environmental contamination, development of insect resistance to pesticides, and disruption of natural control systems.

Management or systems procedures have been advocated for a number of years as more efficient approaches to solving problems in many fields (Ashby, 1963; Bellman, 1957; Quastler, 1965; Thrall *et al.*, 1954; Watt, 1964; Wilde, 1964). Detailed suggestions have also been presented on how integrated management systems might be applied to insect control (Chant, 1964, 1966; Morris, 1963; Nat. Acad. Sci., 1969; Pres. Sci. Adv. Comm., 1965; Smith and van den Bosch, 1967; Watt, 1963, 1964). Despite such readily available

TALL TIMBERS CONFERENCE ON ECOLOGICAL ANIMAL CONTROL BY HABITAT MANAGEMENT, 1970, pp. 83-101.

data, most pest control research programs conducted by universities and governmental agencies have continued to rely on unilateral control methods and show a lack of clear philosophies, priorities, or strategies for dealing with most major pest species. This may appear surprising in view of the fact that the consequences of such unilateral methods with reference to insecticide abuse has been well documented (Newsom, 1967; Pres. Sci. Advis. Comm., 1963; Reynolds, 1969; Ripper, 1956; Smith, 1966; Smith and van den Bosch, 1967; van den Bosch, 1969, van den Bosch *et al.*, 1970). Yet it is not surprising when we consider that most of us—and entomologists are no exception—find it difficult to shift our energies from areas which we have firmly mastered and in which we have achieved recognition to areas that may involve new skills, offer less results in terms of short-term successes, and be fraught with many risks or unknown variables. Nevertheless, this is what is demanded of economic entomologists if they are to respond to the environmental problems that confront their discipline.

DEVELOPING NEW STRATEGIES

A group of us working in the San Joaquin Valley of California are attempting to demonstrate that pest management programs utilizing knowledge and resources currently available are feasible in agricultural crops. Our research is focused specifically on developing a management system for pests of cotton (van den Bosch *et al.*, 1970), which is California's most important farm crop. Cotton represents a particularly challenging crop for developing an integrated pest management program because it reportedly receives more than half of the total insecticides applied to crops in the United States (Eichers *et al.*, 1968). One of two subphases of our research is concentrated on *Heliothis zea* (Boddie), one of this country's three most serious and widespread insect crop pests (Hardwick, 1965), and one of two major cotton pests in California (Reynolds, 1969; van den Bosch, 1969). This paper will be confined to a discussion of how sampling procedures can form the basis for pest management strategies and a report on related research results obtained so far in working on *H. zea*.

Before proceeding further, I should like to define two terms employed in this discussion. By *pest management*, I refer to the bringing together of two or more methods of control into a harmonized system designed to maintain pests below the level of economic damage (Nat. Acad. Sci., 1969). Such a system must necessarily rest on the application of firm ecological principles and approaches. By the term *control*, I refer only to restricting the economic loss caused by a pest population. This is not synonymous with population reduction, although unfortunately both meanings frequently are used without clarification (Smith and van den Bosch, 1967; Turnbull and Chant, 1961).

Basically, all of the pertinent factors that must be considered in selecting control strategies for a pest management system are related to obtaining reliable estimates of pest population density. Thus, the determination of dependable *economic thresholds* (population levels at which supplemental control measures are justified) are primarily subject to the reliability of the sampling methods used to estimate pest populations. If pest populations are overestimated, unnecessary insecticide treatments or other control measures will be employed and costs to the grower may prove greater than any gain derived from increased yields.

Unfortunately, invalid estimates of economic thresholds are more often the rule than the exception in agriculture. The majority of such estimates are not derived from experimental data, but are based instead on traditional practices, arbitrary selection, estimates of population levels used in other areas, and similar nebulous criteria. If data are not available for determination of economic thresholds, growers should be so informed. On the contrary, however, the presence of even unjustifiably low population densities of a pest usually results in a decision to treat crops despite a lack of sampling data. Such decisions are based on the belief that pest populations are unpredictable and may increase to damaging levels before control measures can be effected. Actually, such an "overkill" philosophy stems from fear based on ignorance: a lack of methods permitting confident assessment or prediction of population levels and insufficient knowledge of the potential damage various pest densities may or may not cause. There is a critical need for economic threshold data based on reliable pest population estimates in areas of intense competition between chemical insecticide salesmen. Development of reliable data collection procedures will make it possible to convince growers and farm advisors that higher densities of pests can be tolerated in a crop than have formerly seemed advisable and that these are not necessarily responsible for reduced yields (González *et al.*, 1967).

In our San Joaquin Valley studies, we are striving to develop pest management programs based on reliable economic thresholds, since this will permit greater utilization of the "field insectory" (the efficient predators and parasites found abundantly in and around a crop). In practice, reliable economic threshold data usually results in reduced insecticide treatments. On extensive crops such as cotton, reduced treatments lead to greater area-wide abundance of the endemic and migratory beneficial populations, which in turn serve to regulate pest populations in a variety of adjacent crops. Preservation of the beneficial insects should therefore be a major economic consideration whenever chemical control is contemplated.

Once wide-scale insecticide treatments are begun in a given area, however, subsequent follow-up treatments usually become necessary to control potentially greater numbers and a greater diversity of pest insects. These numbers may be from *resurging* populations (target pests building up in the absence of natural enemies); they may be *secondary* pests (insects not normally pests but with a biological capability to readily become so in the absence of natural enemies); or they may be from *resistant* populations (those insects no longer killed by the chemical treatments applied). Other problems that may result from excessive or unjustified use of insecticides include excessive accumulation of residues, health hazards, legal complications, and the destruction of other beneficial species, including soil conditioning arthropods, pollinating insects, and wildlife. Finally, there is the increased expense of employing insecticides, which involves recurrent costs for equipment, labor, and material. The severity of such problems is documented in greater detail elsewhere (Nat. Acad. Sci., 1969; Newsom, 1967; Ripper, 1956; Reynolds,

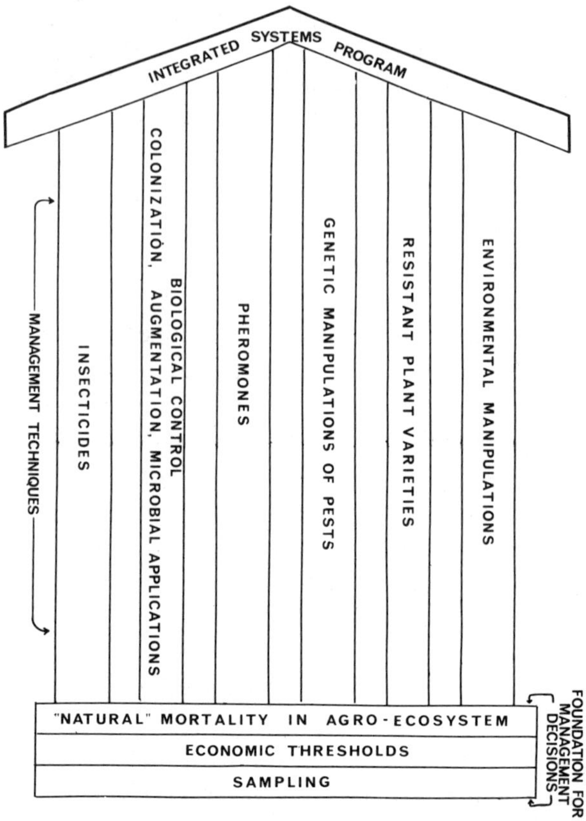

Fig. 1. A schematic diagram depicting development of a pest management program analagous to building a house.

1969; Smith and van den Bosch, 1967; van den Bosch, 1969; van den Bosch *et al.*, 1970).

If advantage is taken of the mortality factors already operational in an agro-ecosystem, however, we believe that supplemental control methods can be devised more efficiently in terms of net optimal crop production relative to pest population levels. Supplemental methods can be thought of as the building blocks of the program, analogous to the supporting beams of a house as illustrated in Figure 1. The foundation consists of reliable sampling procedures that make it possible to determine economic thresholds and the natural mortality in the agro-ecosystem. The beams of the house (the multi-tier of columns supporting the roof in Figure 1) represent potential control methods that can be combined into an integrated pest management program (the roof). Increasing the number of workable control methods generally increases the stability of the interactions between a pest species and its environment, thus maintaining a relatively lower population level of pests.

Our approach to developing a pest management program for *H. zea* involves (1) developing reliable sampling procedures; (2) analyzing the dynamics of the pest population in terms of natural mortality factors present in the agro-ecosystem; (3) obtaining data that permits prediction of future population trends so that factors influencing these trends can be adjusted through modifications or manipulations of the environment; (4) obtaining data on the economics of crop production relative to this pest species and its control; and (5) testing the effectiveness of control methods to determine those most suitable for incorporation into a pest management program. A more detailed discussion of the general principles involved in developing new control strategies may be found elsewhere (Chant, 1964, 1966; Nat. Acad. Sci., 1969; Smith and van den Bosch, 1967).

PROGRAM METHODS

Development of a pest management program for *H. zea* in the San Joaquin Valley requires long-term research to achieve our goal. Nevertheless, considerable progress has already been made and the results obtained have demonstrated the significance of sampling procedures as a basis for developing pest management strategies.

Our research strategy involves a team approach with four entomologists assisted by a farm advisor directing a series of broad studies.[1] Significant contributions to the broad areas of investiga-

[1] Projects are under the direction of the following: L. A. Falcon, University of California, Berkeley; T. F. Leigh, University of California, Davis; D. Gonzalez, University of California, Riverside; R. van den Bosch, University of California, Berkeley; and J. H. Black, Farm Advisor, University of California, Kern County.

tion are also being made by graduate students who are given an opportunity to select highly specialized areas of investigation within the major research projects.

Field studies are being carried out in replicated 80-acre block fields of San Joaquin Valley cotton growers under customary conditions of planting, fertilization, and irrigation. Since minimum effective plot size for any experiment was found to be about 12 acres, the various experiments to date have all been on plots of at least 12 acres or larger. Smaller plots were found difficult to evaluate because of problems created by insecticide drift and interplot movement of adult predators. Figure 2 presents a schematic diagram of the plot arrangement for a typical 80-acre block experiment. Light traps placed outside the fields were employed only for the insecticide-treated and untreated plots.

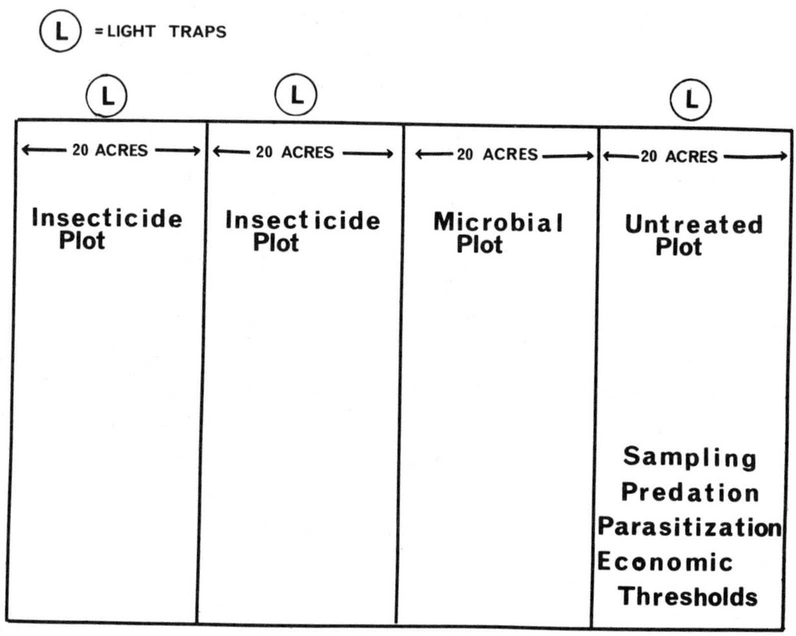

Fig. 2. Schematic diagram of field plot arrangement in team approach to pest management program for *H. zea*.

Our emphasis to date in the University of California, Riverside, project has been on gathering data on the dynamics of *H. zea* on cotton through development of a reliable sampling procedure and the measurement of the effects of predators, parasites, and other obvious "natural" mortality factors against larvae present in the cotton fields as they relate to optimum economic thresholds. The experimental design for this study has been presented by González *et al.* (1967).

Although sampling procedures are still being developed, sufficient data are already available to make the following generalizations. The spatial pattern of *H. zea* larvae within our study areas appears to be nearly random at low infestation levels (i.e., up to a density of approximately 14 larvae less than ½ in. in length) in the fruiting parts on the upper halves of 100 plants. This pattern is indicated by comparison of the fit of our untransformed data to a line for data fitting a theoretical Poisson distribution in which the variance equals the mean (Fig. 3). The X's in the graph represent points significantly different from a Poisson distribution as analyzed by the Index of Dispersion (Grieg-Smith, 1964). On the same graph, the variance for the field data is shown to frequently be larger than the mean at higher densities (beginning at about 16 larvae per 100 plants), indicating a shift to a contagious statistical distribution (Grieg-Smith, 1964). This suggests a gradual change to clumped or aggregated spatial patterns of larvae at higher field densities. Southwood (1966) presents a useful summary of the approaches used in various studies of similar data. In our case, most decisions regarding supplemental control measures are made at densities having spatial patterns overlapping between the nearly random and slightly clumped.

Our sampling procedure thus represents somewhat of a compromise between a random and stratified method. In essence, we have a nested (or hierarchical) design in which we stratify areas within each plot and then draw sub-samples at random from different crop rows within each designated area. We restricted random samples to different crop rows because our data indicated that there was much greater variation in larval counts between rows

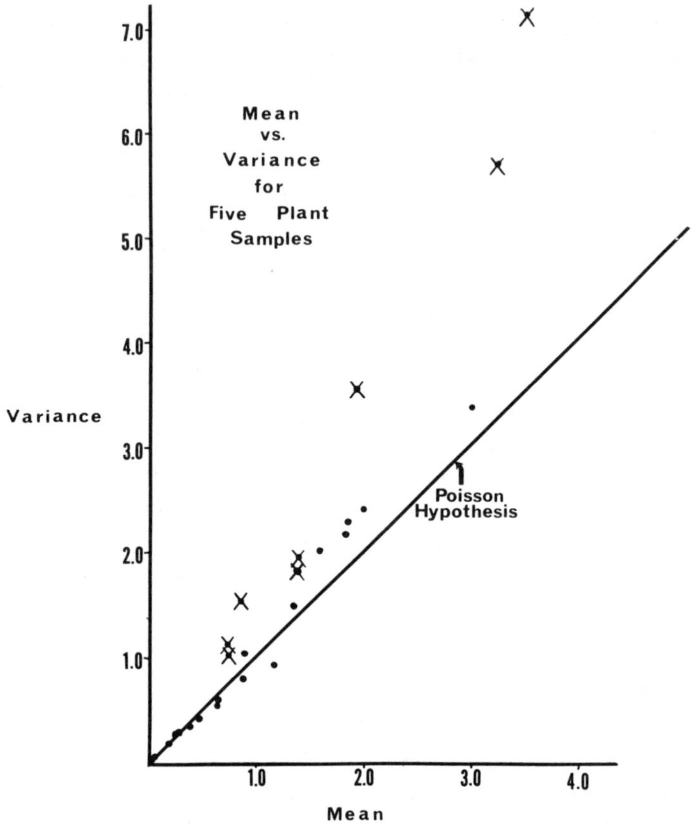

Fig. 3. A theoretical line for a Poisson distribution versus points obtained from larval counts obtained from 2 untreated fields.

than along rows (suggesting that moths may prefer to fly along rather than across rows). We further restricted our search on each plant to the fruiting bodies on upper halves of cotton plants because a preponderance of surviving larvae were found in those locations (Gonzalez *et al.*, 1967). These general conclusions on the best sampling procedure to employ were based on an analysis of counts made over three years in six fields. In addition to giving us a more reliable estimate of populations, this approach has resulted in a more efficient sampling procedure since larger numbers of the sites most likely to have larvae can be examined because of the time saved in avoiding a strictly random sampling procedure for a non-randomly dispersed population.

Mortality of *H. zea* caused by parasites, predators, and disease was assessed in the plots and in the larvae used in the sampling studies. No evidence of significant occurrence of disease was found. Egg and larval parasitization by native parasites was very low with maximum levels infrequently reaching 2 to 3 percent. Parasitization is thus apparently not a key mortality factor (Morris, 1959), particularly since all parasites found to date have been polyphagous (reported on a wide range of lepidopterous hosts). Consequently, we have recently initiated a foreign exploration program to seek more effective egg and larval parasites of *H. zea*.

Studies to determine the major predators of *H. zea* and their searching and feeding capacities have only recently been initiated (Leigh *et al.*, 1970; van den Bosch *et al.*, 1969). Results of preliminary investigations indicate that some predators can cause substantial mortality to *H. zea*, particularly during its early larval instar stage.

Data collected for development of economic thresholds over a 3 year period has indicated no direct correlation between various levels of bollworm infestations and yields in untreated plots varying in size from 12-20 acres. This has been the case regardless of the number of fields involved for worms up to ½ in. in length and for total worms of all sizes. Numbers of worms found on individual dates were added cumulatively to obtain different totals by combining worm counts from different weeks. Counts of the numbers of worms from 4 weeks (minimum) to 10 weeks (maximum) were used in our analysis and were not statistically significant when compared with yields in corresponding plots (Fig. 4). In summary, all data collected from our untreated plots suggest strongly that relatively large numbers of *H. zea* larvae can be tolerated by cotton without economic loss. The data further suggest that variation in yield is primarily due to factors other than feeding by *H. zea* larval populations. The most obvious suspects for such yield reduction are weather, types of soil, and cultural practices.

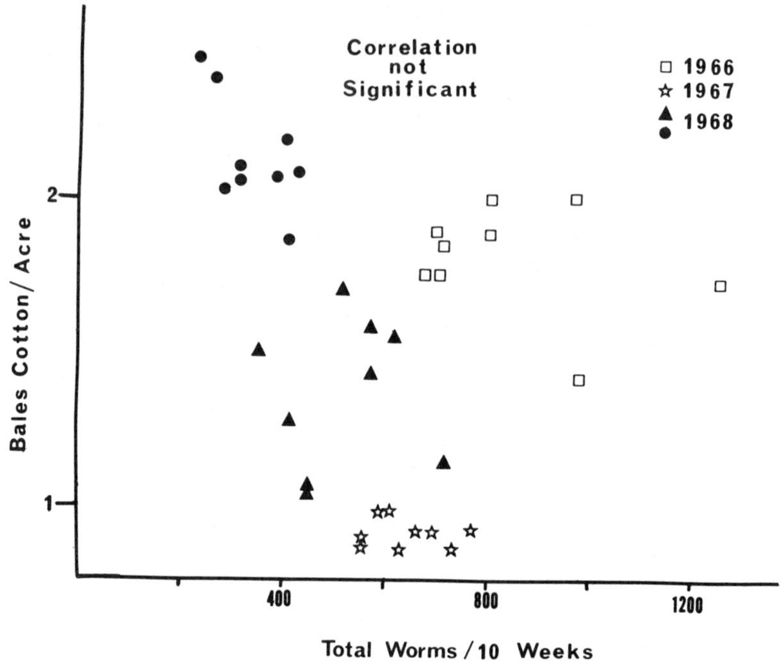

Fig. 4. Scatter diagram for yields versus larvae from untreated plots replicated 9 times/farm.

In the past 2 years, comparisons of yields from untreated large plots versus yields in plots treated with two insecticides and *Bacillus thuringiensis* Berliner (BT) have further indicated that our relatively high larval population levels did not significantly affect crop yields (Table 1). In fact, all but one of seven treated plots yielded less cotton than the untreated standard. Furthermore, over a 2 year period, untreated plants in all cases yielded over two bales of lint cotton per acre even while sustaining what are considered traditionally to be devastating infestations of *H. zea* larvae. In addition to greatly increasing production costs, none of the treatments have been particularly effective in reducing larval populations of *H. zea*. At best a larval population reduction of 47 percent was obtained, and this was achieved only after multiple applications (mean of 7 or more) of methyl parathion. In contrast, mortality levels

Table 1. Summary[a] of treatments used in attempting to develop an economic threshold for cotton bollworm in California.

Year	Material	Number of Applications	Total Larvae/ 10 Weeks	Lbs. Lint Cotton/ Acre
1968	Methyl Parathion (LT)[b]	8	152	1223
	Sevin (LT)	6	193	1270
	Methyl Parathion (UC)[c]	4	202	1608
	Untreated	0	287	1463
1969	Methyl Parathion (UC)	7	272	949
	Sevin (8)[d]	8	306	847
	Sevin (UC)	5	389	1025
	Bacillus thuriengiensis	9	411	1021
	Untreated	0	517	1071

[a] Means from 3 farms 1968, 5 farms 1969.
[b] Treatments based on prolonged light trap catches of *H. zea* adults.
[c] Treatments based on University of California recommendations. In 1968 and 1969 the treatment level was 15 or more larvae < ½ in. in length in the fruiting parts on the upper halves of 100 plants.
[d] Treatment based on 8 or more larvae < ½ in. in length in the fruiting parts on the upper halves of 100 plants.

of 66 to 96 percent were reported for *H. zea* larvae in caged tests with naturally occurring predators (van den Bosch *et al.*, 1969). In economic terms, plots treated with insecticides giving the best population reduction provided poor control of economic damage. In both years, such treated plots produced less cotton than those left untreated. The chemical insecticides employed also have been highly destructive to the "field insectary" of beneficial insects, an already available, cost-free, and potentially effective source of mortality, which when once destroyed is usually not regained during that season.

Our results in terms of economic threshold levels differ appreciably from those reported by Adkisson *et al.* (1964a, b), but our experimental conditions also varied greatly from theirs in terms of plot size and possibly with regard to predator population levels. In contrast to the relatively low economic threshold levels suggested in their studies, our economic threshold in California of 20 larvae less than ½ in. in length in fruiting bodies on 100 upper plant halves is at best conservative (i.e., a low estimate as evidenced by the yield data presented here). Added support for our threshold, however, comes from the knowledge that the high yield data from untreated plots were obtained in a cotton-growing area of the San Joaquin Valley that consistently sustains the highest levels of *H. zea* larval infestation. In all cases, our plots were characterized by unusually high larval population densities. Procedures for sampling and determining economic thresholds of *H. zea* in California cotton are described in detail in our bulletin on a pest and disease control program for cotton (Akisson *et al.*, 1970).

Another aspect of our continuing research program involves an analysis of the relationships between various levels of temperature, light, and humidity and adult fecundity, survival, and egg viability of *H. zea*, data that can ultimately be related to prediction of larval infestation levels. Such more narrowly oriented laboratory and field studies are being concentrated on by a graduate student as a dissertation problem. His contributions have already demonstrated the value of integrating students into the sort of team approach that is required in developing systems procedures. Preliminary results from this research suggest that temperature may be the most critical factor responsible for *H. zea* larval populations at times reaching damaging levels in San Joaquin Valley cotton fields.

DISCUSSION

I consider our most useful research accomplishment to date to be the development of a preliminary sampling procedure for estimating densities of *H. zea* larval populations. Although inefficient in terms of the relatively large sample size presently required, this procedure has yielded consistent results and added confidence to our population assessments, particularly as regards economic thresholds. The discovery that cotton can tolerate relatively high numbers of *H. zea* larvae under California conditions has provided us with greater latitude in searching for natural mortality factors operative up to this higher level. In short, it means that we have up to 3 or 4 weeks more in any given growing season in which to further manipulate the natural mortality factors.

But the discovery is also of considerable economic significance to cotton growers. The present *H. zea* larval population level at which we suggest treatment with insecticides is now more than 10 times higher than that recommended 5 years ago. In practice, this actually means that chemical treatments are only rarely economically justified. It is now clear that in the San Joaquin Valley, where farmers can usually obtain reasonable returns from untreated cotton, insecticide treatments constitute a high risk investment, not a form of insurance essential for saving a crop. They should be employed by growers only after a careful consideration of anticipated gains versus potential risks.

Our preliminary sampling procedure enabled us to consistently demonstrate also that economically damaging levels of this pest rarely develop in previously untreated large fields. Where damaging levels occurred, they were found only after the beginning of August. Thus, it is evident that early season treatments of *H. zea* are not particularly justified since they bring no economic return and in fact can cause financial loss in decreased yields through aggravation of pest problems and an adverse effect on fruiting cotton plants. Apparently, adverse effects from insecticides on the plants are most pronounced during the peak bloom which for the Acala SJ-1 variety occurs around the third week of July in the San Joaquin Valley. With one exception, unharvested green bolls remained in the field at the end of the season in our insecticide-treated plots and yields were substantially reduced. These results differ considerably from those of Adkisson *et al.* (1962), who reported that various combinations of insecticides had no adverse effects on plant growth or fruiting.

In respect to late season (after August) control treatments in the San Joaquin Valley, the preference of young *H. zea* larvae for squares and tender terminals precludes their importance as an economic pest. Essentially, these larvae feed on plant parts that are not harvested. A more detailed discussion on a related phenomenon is

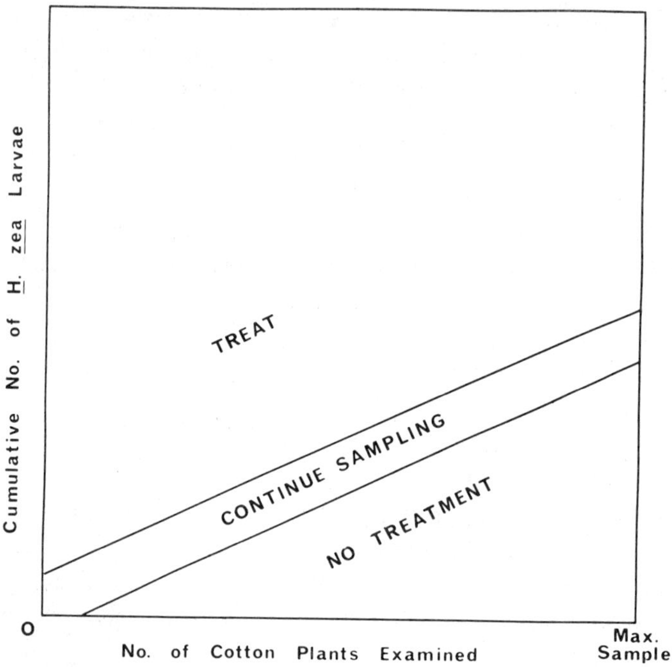

Fig. 5. Schematic sequential graph for *H. zea* larvae on cotton plants.

given by Smith and van den Bosch (1967). Thus, it appears that the critical period for *H. zea* control may be limited to the month of August. If this proves to be correct, it will greatly facilitate the development of supplemental methods of control.

Recently, we have begun developing a sequential sampling plan for *H. zea* larvae on cotton that offers promise of being a highly effective management procedure. Such a plan will emphasize classification of populations in terms of potential economic injury rather than as estimates of population levels (Waters, 1955). For example, under this system a person checking fields will simply record the presence or absence of larvae in each predetermined sample unit as points on a graph (Fig. 5). When a minimum number of cotton plants are examined and a predetermined minimum cumulative larval total is obtained, either a yes or a no decision can be made on the need for treatment. If the maximum number of samples is reached and points in the graph still remain within the area of no decision (continue sampling), then the conservative decision would be to treat the crop. Similar information to that recorded on the graph can be presented in tabular form for more convenient use in the field. This procedure has the advantage of requiring fewer samples, particularly where population levels are either very low or very high.

Waters (1955) reports that the reduction in sampling time obtained by use of the sequential method often exceeds 50 percent. Although there have been several studies on sequential sampling (Morris, 1954; Oakland, 1950, *In* Southwood, 1966; Sylvester and Cox, 1961; Wald, 1947; Waters, 1955; Wolfenbarger and Darroch, 1965), surprisingly little use has been made of the method in decision making on problems caused by agricultural pests.

One reason for the lack of research on what could be such a clearly useful tool for agriculture in decision making is that development of sequential sampling plans require obtaining a substantial amount of data on the spatial pattern of a study species (i.e., whether it is random, clumped, or regular) before reliable decisions can be made on the statistical distribution of data as to whether they are normal, binomial, Poisson, negative binomial, etc. (Ives, 1954; Morris, 1954; Southwood, 1966; Waters, 1955). Other decisions affecting the usefulness of a sequential plan require data on economic thresholds and on the applicability of the sequential plan to a wide spectrum of field situations (e.g., to various levels of population density each of which may have a slightly different statistical distribution). Various levels of population density thus may also have different spatial patterns and may encounter different host plant conditions. Obviously, the development of a sequential sampling plan requires a long-term research effort. Faced with today's promotion pressures to turn out large numbers of publications, few researchers can afford to engage in such time-consuming effort.

In conclusion, we believe that our long-term research will make it possible to develop an effective pest management program for *H. zea* control. Our next projected step in the development of such a program is to obtain more quantitative data defining the key mortality factors affecting *H. zea* on cotton. Because of the relatively simple ecosystem involved, we believe that these factors will be relatively few and will not constitute a major obstacle to our ultimate objective of developing a population model of the major predictive and manipulative components for our system (Chant, 1964, 1966). Concurrently, we will continue to seek effective palliatives that may be used to prevent crop damage until more permanent control measures are found.

We are optimistic about the prospects for developing a management system for *H. zea* in cotton and believe that it is only through such approaches that the quality of our environment can be improved. Hopefully, other agricultural entomologists, including those in administrative capacities, will in the near future begin to scrutinize their research priorities and philosophies and recognize the need for well-rounded ecological approaches to solving environmental problems. Because our environmental resources are being rapidly depleted, we cannot continue to afford the luxury of living in relative ignorance of the means for ensuring their optimum utilization.

ACKNOWLEDGEMENTS

The work described in this paper represents the input of many entomologists, a majority of them graduate students. In addition to the project leaders mentioned earlier, I wish to recognize the contributions of J. Allen (UCR), T. Ashley (UCR), S. Davis (UCB), L. Ehler (UCB), K. Eveleens (UCB), J. Galagher (UCB), C. Jackson (UCD), C. Olivares (UCR), G. Orphanides (UCR), and D. Ramsey (UCR). Señores J. Allen (UCR), and C. Huzar, Department of Statistics (UCR), provided the statistical advice essential to this research.

Although the ideas presented are the fruitful products of discussions among the above named individuals and consultations with Señor José Everto Dietrick and many others too numerous to mention, I accept full responsibility for my interpretation of the data.

I am particularly grateful to Señor H. W. Lawton for his assistance in editing this manuscript.

The study was made possible by partial financial support from the USDA Cooperative State Research Grant 816-15-14, the Cotton Producers Institute, and California Planting Cotton Seed Distributors.

LITERATURE CITED

Adkisson, P. L., J. Hacskaylo, R. L. Hanna, and J. K. Walker, Jr. 1962. Flowering rate, boll maturity, and yield of cotton treated with various insecticides. Prog. Rept. 2245. 7 p. Texas Agr. Exp. Sta.

Adkisson, P. L., R. L. Hanna, and C. F. Bailey. 1964a. Estimates of the numbers of *Heliothis* larvae per acre in cotton and their relation to the fruiting cycle and yield of the host. J. Econ. Entomol. 57:657-63.

Adkisson, P. L., C. F. Bailey, and R. L. Hanna. 1964b. Effect of the bollworm, *Heliothis zea*, on yield and quality of cotton. J. Econ. Entomol. 57:448-50.

Akisson, N. B. et al. 1970. Pest and disease control program for cotton. Calif. Agr. Exp. Sta. Berkeley, 20 p.

Ashby, W. R. 1963. An introduction to Cybernetics. Science Editions. Wiley, N. Y. 295 p.

Bellman, R. 1957. Dynamic Programming. Princeton University Press, Princeton, N. J. 342 p.

Chant, D. A. 1964. Strategy and tactics of insect control. Can. Entomol. 96:182-201.

Chant, D. A. 1966. Integrated control systems. p. 193-218. *In* Scientific aspects of pest control. Nat. Acad. Sci.—Nat. Res. Counc. Publ. 1402. Washington, D. C. 470 p.

Eichers, Theodore, Paul Andrilenas, Robert Jenkins, and Austin Fox. 1968. Quantities of pesticides used by farmers in 1964. Agr. Econ. Rept. No. 131. Econ. Res. Serv. USDA. Washington, D. C. 37 p.

Gonzalez, D., R. van den Bosch, G. Orphanides, L. Dawson, and C. White. 1967. Population assessment of cotton bollworm in relation to pest control practices. Calif. Agr. 21(5):12-14.

Greig-Smith, P. 1964. Quantitative plant ecology (2nd ed.) Butterworths. London. 256 p.

Hardwick, David F. 1965. The corn earworm complex. Ent. Soc. Can. Mem. 40. 247 p.

Ives, W. G. H. 1954. Sequential sampling of insect populations. Forestry Chron. 30:287-91.

Leigh, T. F., D. Gonzalez, and R. van den Bosch. 1970. A sampling device for estimating absolute insect populations in cotton. J. Econ. Entomol. 63:1704-06.

Morris, R. F. 1954. A sequential sampling technique for spruce budworm egg surveys. Can. J. Zool. 32:302-13.

———. 1959. Single factor analysis in population dynamics. Ecology 40:580-88.

———. 1963. (ed.) The dynamics of epidemic spruce budworm populations. Mem. Ent. Soc. Can. 31. 332 p.

National Academy of Sciences. 1969. Integrated systems of pest management. p. 447-82. *In* Insect pest management and control. Principles of plant and animal pest control. Vol. 3. Nat. Acad. of Sci. Washington, D. C. 508 p.

Newsom, L. D. 1967. Consequences of insecticide use on non-target organisms. Ann. Rev. Ent. 12:257-86.

Oakland, G. B. 1950. An application of sequential analysis to whitefish sampling. Biometrics. 6:59-67.

President's Science Advisory Committee. 1963. Use of pesticides. Life Sciences Panel. The White House. U. S. Govt. Printing Office, Washington, D. C. 25 p.

———. 1965. Restoring the quality of our environment. Report of the environmental pollution panel. The White House. U. S. Govt. Printing Office, Washington, D. C. 317 p.

Quastler, H. 1965. General principles of systems analysis. P. 313-33. *In* Theoretical and Mathematical Biology. Blaisdell Pub. Co., New York. 418 p.

Reynolds, H. T. 1969. Current pest control situation on cotton in the United States with particular reference to California. P. 12-18. *In* Proc. Cotton Symposium. University Calif., Div. Agric. Sciences, Berkeley, 146 p.

Ripper, W. E. 1956. Effects of pesticides on balance of arthropod populations. Ann. Rev. Ent. 1:403-38.

Smith, E. H. 1966. Advances, problems, and future of insect control. P. 41-72. *In* Scientific aspects of pest control. Nat. Acad. Sci.—Nat. Res. Counc. Publ. 1402. Washington, D. C. 470 p.

Smith, R. F. and R. van den Bosch. 1967. Integrated control. P. 295-340. *In* W. W. Kilgore and R. L. Doutt, eds., Pest control. Biological Physical and Selected Chemical Methods. Academic Press, New York. 477 p.

Southwood, T. R. E. 1966. Ecological methods. Methuen and Co., Ltd., London. 391 p.

Sylvester, E. S., and E. L. Cox. 1961. Sequential plans for sampling aphids on sugar beets in Kern County, California. J. Econ. Ent. 54:1080-84.

Thrall, R. M., C. H. Coombs, and R. L. Davis, eds. 1954. Decision Processes. Wiley, N. Y. 332 p.

Turnbull, A. L., and D. A. Chant. 1961. The practice and theory of biological control of insects in Canada. Can. J. Zool. 39:697-753.

van den Bosch, R. 1969. A comprehensive analysis of the University of California Cotton Insect Research Program. P. 98-112. *In* Proc. Cotton Symp. Insect and Mite Control Problems and Research in California. Univ. Calif., Div. Agric. Sci. Berkeley, 146 p.

van den Bosch, R., T. F. Leigh, D. Gonzalez, and R. E. Stinner. 1969. Cage studies on predators of the bollworm in Cotton. J. Econ. Entomol. 62:1486-89.

van den Bosch, R., T. F. Leigh, L. A. Falcon, V. M. Stern, D. Gonzalez, and K. S. Hagen. (In press). The developing program of integrated control of cotton pests in California. Chapt. 17 *in* Theory and Practice of Biological Control. Plenum Press, New York.

Wald, A. 1947. Sequential analysis. John Wiley & Sons, New York. 212 p.

Waters, W. E. 1955. Sequential sampling in forest insect surveys. Forest Science. 1(1):68-79.

Watt, K. E. F. 1963. Mathematical population models for five agricultural crop pests. P. 83-91. *In* E. J. LeRoux (ed.) Population dynamics of agricultural and forest insect pests. Ent. Soc. Can. Mem. 32.

———. 1964. The use of mathematics and computers to determine optimal strategy and tactics for a given insect problem. Can. Entomol. 96:202-20.

Wilde, D. J. 1964. Optimum seeking methods. Prentice-Hall, Inc. Englewood Cliffs, N. J. 202 p.

Wolfenbarger, D. A., and J. G. Darroch. 1965. A sequential sampling plan for determining the status of corn earworm control in sweet corn. J. Econ. Entomol. 58:651-54.

PEST MANAGEMENT SYSTMES

Integrated Systems of Pest Management

Subcommittee on Insect Pests
National Academy of Sciences

Much of the material discussed in previous chapters has dealt with the principles involved in various individual methods of insect-pest control and management and with the tools and techniques available for making the methods effective. Each control method has an important place, but none has always provided a satisfactory solution to the many problems posed by insects and other arthropod pests. Dramatic successes can be cited for the different control methods, but equally dramatic failures have occurred. Problems have accelerated during the last two decades that demonstrate conclusively that single-factor, *ad hoc* approaches to the control of insects and related pests are inadequate.

An attitude of unreserved optimism became prevalent among most entomologists with demonstration of the spectacular effectiveness of DDT. Failures of synthetic organic insecticides to control all pests, and adverse consequences of their widespread use during the last two decades, have changed this attitude to a more rational but somewhat pessimistic one. Development of insecticide-resistant populations, resurgence of treated populations, elevation of secondary pests (or in some cases, previously innocuous species) to a status of primary importance, deleterious effects on populations of nontarget organisms, and general pollution of the environment with measurable residues of persistent chemicals pose increasingly critical problems associated with the use of synthetic organic insecticides.

Attempts to control pests by biological means have resulted in striking successes in some cases. Importation and dissemination of exotic parasitic and predacious insects, development and use of insect-resistant plant varieties, and dissemination of insect pathogens have given excellent control of a few species.

PRINCIPLES OF PLANT AND ANIMAL PEST CONTROL, Vol. 3: INSECT-PEST MANAGEMENT AND CONTROL, Publication 1695, Committee on Plant and Animal Pests, Agricultural Board, National Academy of Sciences—National Research Council, Washington, D.C. 1969.

There are many insect problems, however, that cannot be solved by any one of these approaches.

The rapidly evolving problem of controlling insects and related pests suggests the probability of increasing difficulty. The unprecedented rate of increase of the the world population imposes the necessity for greater production to meet man's needs. Efforts to feed and clothe a population expected to double by the end of the twentieth century will force agriculture toward the comparative simplicity of monocultures, with their tendency toward encouraging violent oscillations in pest populations. Problems posed by such genetically plastic and evolutionarily resourceful organisms as insects and related arthropods and their interactions with a rapid changing agriculture are too dynamic to be solved by reliance on one-component control systems. Obviously, man is not making the best use of his ability to control insects in view of what is known about the factors governing the fluctuations of populations in nature. Man's efforts to impose his will on nature have amounted, until recently, to little more than what has been termed "bulldozing nature," with little thought to consequences. His efforts to control arthropod pests have sometimes resulted in as much harm as good. During the last few decades, substantial information has accumulated to suggest that pest control must be extended beyond empirical methods to a system based on principles of applied ecology. Such a system has been developing rapidly; it has come to be known as integrated control.

DEFINITION OF INTEGRATED CONTROL

Several definitions of integrated control have been proposed. They include:

1. Applied pest control that combines and integrates biological and chemical measures into a single unified pest-control program. Chemical control is used only where and when necessary, and in a manner that is least disruptive to beneficial regulating factors of the environment. It may make use of naturally occurring insect parasites, predators, and pathogens, as well as those biotic agents artificially increased or introduced.
2. An ecological approach to pest management in which all available necessary techniques are consolidated into a unified program, so that populations can be managed in such a manner that economic damage is avoided and adverse side effects are minimized.
3. A program of arthropod-population management designed to keep pest populations below economic tolerance levels by maximizing environmental resistance and supplementing this by use of selective pesticide applications if economic tolerance levels are threatened.
4. Utilization of all suitable techniques to reduce and maintain pest populations at levels below those causing injury of economic importance to agricul-

ture and forestry, or bringing together two or more methods of control into a harmonized system designed to maintain pests at levels below those at which they cause harm—a system that must rest on firm ecological principles and approaches.

All the proposed definitions have one common theme: the system must be based on sound ecological principles. Early definitions emphasized a two-component system based on biological and chemical control, and the concept of integrated control has come to be most frequently thought of as a blending of these two control techniques.

HISTORY OF INTEGRATED CONTROL CONCEPT

Integrated control is a new name for a philosophy and system of pest control that entomologists, particularly in forestry, have been using for years. When analyzed, successful control programs attributed to single factors are often found to have involved factors other than those deliberately invoked. Integrated control differs from past approaches in being a deliberately conscious attempt to blend and harmonize the different suitable techniques making up the control system to solve pest problems better.

Integrated control, as thought of today, began about 25 years ago in widely separated places: Germany, Nova Scotia, and California. Pickett and his co-workers in Nova Scotia stressed their conviction that ecological concepts and methods must be applied to the problems of controlling insects. By this they meant that a thorough understanding of the dynamics and ecology of pest populations would lead to a method of control that would rely not only on pesticides, but also on natural enemies. They advocated the development and use of spray programs that would have minimal harmful effects on beneficial species, which could then exert their full effects on the pest. They called this approach the modified program but eventually substituted integrated control, a term apparently first proposed by Bartlett in California in 1956. It is based on achieving maximum effectiveness of natural control agents already established in an ecosystem (rather than on introducing and disseminating exotic species) and in using, in addition, selective pesticide applications when necessary. At one time the program was being used on more than 80% of the apple and pear acreage in Nova Scotia, with notable success and drastic reductions in the cost of pest control. However, as a result of radical changes in economics and increased demand for blemish-free fruit, growers in the area have reverted to the traditional method of chemical control.

Integrated control in California owes its origin to the successful supervised control procedures worked out for alfalfa and cotton in the late 1940's. With supervised control, growers arrange for entomologists to check fields carefully and frequently and to recommend the application of pesticides only when pest populations are found to be increasing beyond the restraint of natural mortality factors in the environment.

A similar type of supervised control of forest insect pests had been practiced in Germany for more than 25 years. No control operations were started until the role of natural mortality factors had been carefully evaluated.The comparative sophistication of the supervised control program for control of forest pests is demonstrated by the fact that the first life tables for insects were published 35 years ago.

Coincident with these advances, much interest has been shown in the desirability of applying ecological principles to problems of controlling or regulating pests of importance to agriculture and forestry. This interest has contributed to clarification of thinking on the principles of pest control and to more precision and sophistication in the definition of objectives. The following points have been emphasized: (1) there is need to study the ecosystem as a whole, and pest control must become applied ecology; (2) the best methods of controlling pests can be devised only with a knowledge of the principles underlying the fluctuations in pest populations; (3) the control of populations is a function of the ecosystem, and a knowledge of all the principal elements of the system is essential to an understanding of population phenomena; and (4) fundamental work on the ecology of a pest, combined with studies of control methods, is ultimately more fruitful in reducing damage than the purely *ad hoc* approach.

EVOLUTIONARY PROCESS OF PEST MANAGEMENT

It is obvious from early papers that integrated control was developed as a concept of blending chemical and biological control techniques, and many workers still hold to the idea of a two-component system. In the early 1960's the first suggestions arose for broadening the concept to include the integration, not only of chemical and biological control methods, but of all practices, procedures, and techniques relating to crop production, into a single unified program aimed at holding pests at subeconomic levels. Thus, current thinking has evolved from the relatively narrow concepts of the two-component system—chemical and biological control—to the much broader concept of pest management.

Although integrated control by definition involves at least two components, some of the most successful efforts rest on only one factor. Two classic examples are control of the cottony-cushion scale, *Icerya purchasi* Maskell, by the vedalia, *Rodolia cardinalis* (Mulsant), and control of the grape phylloxera, *Phylloxera vitifoliae* (Fitch), by use of American resistant rootstocks with European grapevines, *Vitis* spp. Neither of these methods, the first biological control and the second control by use of plant resistance, has required the integration of any other technique to give outstandingly successful control of two important pests.

The concept of eradication of pest species from large areas of the world has gained much momentum during the last decade. It clashes with the concept of integrated control, i.e., a blending of biological and chemical methods. The difficulties of reconciling these two approaches are apparent from the intrinsic nature of biotic regulating mechanisms. The maintenance of such mechanisms depends on a continuous supply of hosts and proper environmental conditions. In general, the higher the economic level, the greater the possibility that biotic regulating mechanisms can come into play, and therefore the greater the opportunity for integrated control. In some pest-control situations the population levels that can be tolerated are extremely low or may even be zero. Satisfactory control is very difficult to achieve in any such case, and the very low economic levels usually create impossible goals for biological control and consequently for integrated control.

The necessity for utilizing economic levels of infestation as one of the basic principles of integrated control illustrates the conceptual conflict between two major schools of entomological thinking. One school adheres to pest control from an ecological basis, relies heavily on the two-component system (biological and chemical methods), and accepts the presence of subeconomic populations of pests as being prerequisite to the successful operation of such a system. The presence of pests in populations that are of subeconomic importance are thought of as being analogous to gasoline in an internal combustion engine. It is the fuel that provides the energy for operation of the system; without it the machine cannot function. It is this obvious dependence on the presence of subeconomic levels of pest populations for the successful operation of integrated control systems that forms the basis for the opinion of many entomologists that the concept of eradication is the antithesis of the concept of integrated control.

The philosophy of the second school is essentially the same as that of a great majority of modern farmers. It believes that the presence of subeconomic pest populations is undesirable, and that many pests, if not all, should be eradicated by the invocation of special techniques, or, if not eradicated, kept at the lowest possible levels by regularly scheduled applications of broad-spectrum insecticides.

Although eradication of a species is a concept as difficult to accept by those oriented toward the multidimensional concent of integrated control as is the necessity for maintenance of subeconomic infestation levels by many entomologists, it is recognized as a valid objective and reasonable goal in certain cases. Well-organized and properly executed eradication programs, armed with the necessary tools and proceeding from a base of thorough knowledge of the pest and its ecosystem, provide the ultimate solution to the problem of a number of noxious species. The successful program for eradication of the screw-worm, *Cochliomyia hominivorax* (Coquerel), is a classic example.

There is nothing ecologically abhorrent about the eradication of a species. The geological record shows that many species have become extinct. Some entomologists argue that it is morally wrong for man to attempt to eradicate something that he cannot create. Yet they do not deny man's right to attempt to control insect populations. In either case, the environment must be changed. Eradication projects are directed against pests that otherwise would be the target of other methods of pest control, and eradication may be the lesser of two evils in the ecological sense.

Perhaps a great deal of resistance to the eradication concept results from knowledge that many eradication attempts in the United States have failed and only a few have succeeded. Attempts to eradicate firmly established species by the application of broad-spectrum insecticides at high rates of application to huge areas have usually been unsuccessful. The consequences of such efforts in terms of adverse effects on nontarget organisms have often been detrimental. However, new concepts and improved tools of considerably greater sophistication than any previously known are receiving the attention of entomologists and other researchers. Prospects for their successful use justify a re-evaluation of the potential of eradication techniques for control of an invading species or of a species making significant extensions of its range.

The broad concept of integrated control that has evolved—a concept that includes use of all appropriate techniques, aptly described as a group of devices forming a network for serving a common purpose—allows reconciliation of the differences in the two philosophies. Integrated control systems by definition may lead in either of two directions—eradication or pest management. A combination of methods, acting concurrently or in sequence, may be used to reduce a pest population to very low levels, where special techniques, such as the use of sterility, may eradicate the population. More often, however, an integrated control system will reduce the pest to an acceptable level and then maintain it there—true management of pest populations.

Recently, the term "pest management" has been advanced as one more expressive of current thinking than "integrated control" as a reference to pest control from ecological bases. The term is much more inclusive than integrated control, and it avoids the conflict between proponents of integrated control systems that require the presence of subeconomic levels of pest populations and those who maintain that eradication measures can be harmonized with such a system. Pest management emphasizes the comprehensive nature of the approach and its reliance on ecological principles; it also conveys the idea of the intelligent manipulation of nature for man's benefit—the same kind of manipulation that is involved in fisheries and wildlife management and in the whole system of modern agriculture. Pest management, like integrated control, implies the continued existence of potentially harmful species at tolerable levels of abundance. However, it does not, by definition, imply that pests should not be eradicated whenever feasible, since the extermination of a pest species is likely to be the only completely satisfactory solution to many important insect problems.

The conceptual basis for this thesis rests largely on the principle that relative scarcity is more characteristic of phytophagous organisms than is great abundance, and that pest status is achieved as a result of man's activities. Man's upsets of long-established patterns of intraspecific and interspecific relationships are pest-inducing. Some of the most important and obvious of these activities are: (1) introduction of actual and potential pests to previously uninfested areas; (2) introduction of exotic plants and animals to new areas; (3) growing of improved varieties and strains of organisms; (4) intensive land use and practices of monocultures; and (5) simplification of the ecosystem as a result of agricultural and industrial activities.

Pest management is a system that seeks to deal with the abundance of populations by reducing the fitness of the innate qualities of individuals in a population for operative features of the environment. According to this concept, fitness can be reduced by (1) modifying the innate qualities that enable individuals to perform their life functions or (2) altering essential features of the environment to make it no longer able to support large numbers of a population, while leaving the innate qualities of its members unchanged.

Such a concept considers pest management as comparable with crop and wildlife management. It recognizes as a basic premise that it is only for the selfish welfare of the human being that there is any need for management of populations of any organism. Like integrated control, it recognizes that the future solution to pest problems may be in learning to live with pest species rather than eliminating them. It does not exclude single-factor and eradication techniques, and it avoids the semantic difficulties, implicit in the early concept of integrated control, of being a contradiction in terms. There is much to commend the term "pest management" as a description of the goals of economic entomology. If it were adopted, the term "integrated control" would refer only to ways in which various techniques are used to achieve pest management.

NEED FOR INTEGRATED SYSTEMS OF PEST MANAGEMENT

The rapid development of integrated-control techniques and their widespread acceptance during the last two decades have resulted from a growing awareness of the shortcomings of single-factor approaches to pest control. Rachel Carson was not the first to bring to public attention the obvious drawbacks of excessive reliance on and use of chemical pesticides to deal with problems of pest control. Her literary skill and emphasis on the sensational, however, made her criticisms especially effective. It is unfortunate that errors of logic and fact in her book have seriously eroded the impact of her message.

Sole reliance on chemicals for pest control has the following drawbacks: (1) selection of resistance to insecticides in pest populations; (2) resurgence of treated populations; (3) outbreaks of secondary pests; (4) residues, hazards, and legal complications; (5) destruction of beneficial species, including parasites, predators, and pollinating insects; and (6) expense of pesticides, involving recurrent costs for equipment, labor, and material.

Resistance is a problem in evolution; its development in a pest population is evolution in action. Some individuals in almost every population of pests possess more ability than others to detoxify a chemical and render it harmless or to escape its effects through changes in behavior. High levels of resistance to one or more insecticides occur in some isolated populations of normally susceptible species prior to their exposure to the chemicals. Resistance has been observed in about 220 major pests, including species of mites, mosquitoes, house flies, moths, and beetles. The continuing development of insecticide-resistant biotypes in other species is a virtual certainty. With some species the problem has already accelerated to a frantic race by entomologists to develop new pesticides in order to keep just one step ahead of the pests.

Resurgence is a phenomenon with roots in basic ecological principles. The ability of an organism to reproduce and increase is usually variable; in addition to the organism's genetic composition, one of the major factors influencing this ability is its own density. As density increases, competition for food and other requisites increases and reproduction decreases. If the population rises to extreme heights, reproduction may cease and the population may decline precipitously. At low densities, there is little competition, and reproduction is at or near its maximum for the existing environmental conditions. Thus, a pesticide applied to a high pest population may drive it to a low level where reproduction is great, with the result that the population surges back to its original high and damaging proportions. Pesticide destruction of beneficial parasites and predators that normally hold pest populations in check is often an important factor in resurgence. If, in addition to having a higher reproductive capacity at low densities, a pest is simultaneously freed from the restraint imposed on it by natural enemies, its resurgence or flareback will be that much more rapid.

The outbreak of pests other than those against which pesticides are directed is a common phenomenon. For example, the red-banded leaf roller, *Argyrotaenia velutinana* (Walker), increased in importance on apple in Ontario, Canada, from 1929 to 1956, from an insect of casual or even rare occurrence to one of the most serious with which fruit growers had to contend. During this period, spray programs became increasingly complex, starting with a simple program of lead arsenate and ending with very complex schedules of DDT, parathion, TDE, and malathion. The red-banded leaf roller, a secondary pest, rose to a position of primary importance as a consequence of a pesticide program directed against other species. The mechanisms involved in this phenomenon are not certain. It seems clear, however, that a factor of major importance was the destruction of parasites and predators by use of wide-spectrum pesticides.

It must be stressed that the use of chemical insecticides may change the status of pests in the direction of less importance, even to the extent that major noxious species completely lose pest status. Until about 1949, when DDT came into general use in northeastern Canada, the fruit-tree leaf roller, *Archips argyrospilus* (Walker), which had been the most abundant leaf roller on apple, often causing extensive damage, declined as rapidly as the red-banded leaf roller increased. At the same time that the fruit-tree leaf roller disappeared, populations of the oblique-banded leaf roller, *Choristoneura rosaceana* (Harris), and the three-lined leaf roller, *Pandemis limitata* (Robinson), declined precipitously. Similarly, the use of chlorinated hydrocarbon and organophosphorus insecticides for control of the boll weevil, *Anthonomus grandis* Boheman, in the southern United States resulted in a decline of the cotton leafworm, *Alabama argillacea* (Hübner), and the cotton aphid, *Aphis gossypii* Glover, from the status of major pests to that of relative obscurity.

Hazards to man of the widespread use of chemical pesticides have not been adequately evaluated because of insufficient information. However, there is no evidence that the minute traces of pesticides with which man is in daily contact injure or seriously affect him in any way. This is not the case with many species of wildlife, especially fish. There is ample evidence that substantial populations of various wildlife species in many areas have been destroyed or seriously harmed by contact with pesticides. The hazards from misuse of the more toxic chemicals are quite clear.

Cost is one of the most important drawbacks to excessive reliance on chemicals for pest management. Hundreds of millions of dollars are spent on pesticides each year in the United States, but statistics are poor on the value of the crop saved by the use of chemicals. The economics of pest control is discussed fully in Chapter 18.

One of the most pertinent facts about the comparatively high cost of pesticide programs is that substantial reductions can be made in the amount of chemicals being used for pest control without sacrificing efficiency. This can be done by (1) replacement of scheduled applications with applications based on population assessment, and (2) recognition that many crops can tolerate substantial levels of infestation of some pests without economic loss.

The general public, perhaps unfortunately, has been conditioned to demand food products totally devoid of blemishes produced by pests. Growers have been led to expect absence of all pests, which frequently is not required to prevent economic loss. The demand for such unnecessarily high efficiency in pest control has been further conditioned by the action of regulatory agencies in declaring that insects in food are filth or indicators of filth. These attitudes seriously handicap efforts to adopt pest-management systems that do not require excessive application of pesticides. As the advantages of such systems become better known, consumers may alter their attitudes and accept an occasional blemish.

These points illustrate the weakness of pesticides as strategic tools for pest control. It is now necessary to broaden this indictment somewhat and to point out that no single artificial control technique should be relied on exclusively for the regulation of pest populations. Because of their dynamic, adaptable nature, biological controls can sometimes be excluded from this indictment. In many cases, flexible and natural biological controls have proved permanent and highly effective, and further instances of this will undoubtedly be developed in the future.

Components of agricultural and forest ecosystems are continually evolving. A dynamic, multilateral management system for regulating pests in these changing situations is essential. Most one-component systems lack the flexibility necessary for the suppression of dynamic pest populations. Almost invariably, one-component systems lack permanence and in this sense are palliative rather than curative. The empirical approach to pesticide use usually relieves rather than cures; no matter how intelligently pesticides are used, they can never be more than palliatives. Moreover, many one-component systems create unwanted side effects, because they do not adequately recognize other components of the ecosystem and they tend to disrupt rather than to harmonize with them in advantageous ways. Unilateral systems exert pressures on pest populations from narrow bases and facilitate pest reactions by presenting a single plane of activity. Finally, most pest situations involve complexes of pest species, and this adds dimensions and demands normally beyond the scope of any one-component control system.

Although pesticides have been indicted on various valid grounds, the main drawbacks to their exclusive use, which arise from their great effectiveness, are based on the previous statements regarding the unsuitability of unilateral, one-component control systems. However, no individual control technique, and most emphatically this includes pesticides, should be rejected from consideration in integrated control systems. Each control technique has a potential role to play in concert and harmony with the others, and a major technique such as the use of pesticides can be the very heart and core of integrated systems. Chemical pesticides will continue to be one of the most dependable weapons of the entomologist for the foreseeable future. Rapid expansion of agriculture, in developing countries especially, will demand repressive applications of chemicals for pest control until more acceptable alternative techniques can be developed. There are many pest problems for which the use of chemicals provides the only acceptable solution. Contrary to the thinking of some people, the use of pesticides for pest control is not an ecological sin. When their use is approached from the sound base of ecological principles, chemical pesticides provide dependable and valuable tools for the biologist. Their use is indispensable to modern society.

BASIC PRINCIPLES OF THE INTEGRATED-CONTROL CONCEPT

CONSIDERATION OF THE ECOSYSTEM

The first principle of pest control, around which all other principles revolve, is consideration of the ecosystem. An ecosystem may be defined as an area that includes living organisms and nonliving substances interacting to produce an exchange of materials between the living and nonliving parts. This definition is so broad that it has been criticized as being vague, ineffective, and too all-inclusive to be really useful in ecology. Agricultural ecosystems are no less difficult to define satisfactorily, and the very breadth of the definition emphasizes the most basic point of insect control: successful programs that have any hope for providing solutions to problems rather than serving as temporary palliatives must consider pest populations as one component of complex environments.

The ecosystem is here considered to be a unit composed of the total complex of organisms, including the pest species, their competitors, other associates, and their food, shelter, hosts, and culture; weeds and other plants; soil and water and their management; the over-all conditioning environment; and the various agricultural, industrial, and recreational activities of man. Control of populations is a function of the ecosystem, and knowledge of the role of all the principal elements of the unit is essential to an understanding of population phenomena. The most effective system for controlling pests can be devised only after thorough knowledge has been gained of the principles underlying the fluctuations of populations that make up the ecosystem.

The limits of an ecosystem are usually vague. However, the area encompassed must be large enough that the important biotic components are included to the extent that their major activities fall within its limits. For practical purposes, it is what the problem makes it. For example, an agricultural ecosystem will usually include a group of fields, their margins, and certain other types of surrounding areas, such as woods, streams, and weedy, uncultivated areas. It should be considered as a part of a much larger ecosystem. Areas remotely removed from an agricultural ecosystem, considered in the narrow sense, may be important in contributing some biotic components of the restricted agricultural ecosystem. The coccinellid complex, for example, is highly mobile, and extensive movement occurs from crop to crop and area to area, sometimes to and from distant areas of hibernation.

Size, density, and variety of host plants in the ecosystem affect pest density. These are important factors in pest-management concepts, because they can be managed or manipulated to advantage in regulating pest-population density.

Although ecosystems are complex, there is no need to be overwhelmed by the difficulty of acquiring all the data necessary for developing the most sophisticated integrated-control techniques. Much progress can be made by using simple pest-management programs based on limited understanding of components of the ecosystem. Appropriate and successful beginnings can be made by regarding the ecosystem as a complex of overlapping subsystems, each of which, from the standpoint of pest control, can be analyzed separately, with attention concentrated on the key pest species.

The most important factors in the population dynamics of any pest can be determined rapidly by determining the life-cycle stage most responsible for increasing or decreasing numbers of the pest and then pinpointing the causative mortality factors within this stage. This approach is basically simple and can be applied by anyone who has done a simple regression analysis. It is biologically meaningful and provides a useful framework for the development of population models.

Successful application of pest-management systems demands that the types of pest species in each ecosystem be identified. This allows for the synthesis of the most effective control program and gives proper consideration to the differences in techniques that may be used for control of different types of pests.

There are two primary considerations in focusing attention on the pest. First, in any agricultural or forestry ecosystem there are seldom more than a few key pests, and often there is only one. Serious, perennially occurring, persistent pests that dominate control practices because their populations often remain above economic thresholds and thus require continuing applications of pesticides may be defined as key pests. Second, the kind of damage they cause or their abundance determines the status of key pests. The damage potential of the individual pest species is an important consideration. Low numbers of some pests can cause high damage, for example, vectors of plant and animal pathogens, and those, such as ticks, that secrete toxins producing paralysis. Also, key pests often attack the harvested unit of crops. With these, it may be necessary to begin control procedures at relatively low population levels. For example, tomatoes attacked by the tomato fruitworm, *Heliothis zea* (Boddie), may suffer little loss in yield. However, the presence of a single entry site destroys the marketability of the fruit. When the same pest attacks grain sorghum, it primarily reduces the quantity of grain harvested and affects grain quality very little. Thus, considerably higher levels of infestation by this pest can be tolerated in grain sorghum than in tomatoes. In general, pests that transmit pathogens and those that attack the harvested unit of crops impose unusually stringent requirements on a control program, in that populations must be suppressed and maintained at unusually low levels.

Occasional pests, in contrast to key pests, vary in importance. They create problems only in certain areas or during certain years. Such pests are usually under adequate biological or environmental control. However, disruptions or weaknesses in control occasionally occur, and these permit the pest to increase to economic levels. These occasional pests are especially suited to integrated pest-management techniques, and it is against such pests that a system of corrective as opposed to protective treatment has its greatest impact. Also, insect-resistant crop varieties or strains have great utility for control of pests of this category. It is with these pests that means of predicting the occasional occurrence of damaging populations assume special importance. Accurate predictions of their occurrence in numbers above economic levels allow for the application of control measures in a manner that will bring the regulatory factors back into dominance in the shortest possible time.

Potential pests cause no significant damage under current conditions. In attempts to control key and occasional pests, care must be taken not to alter conditions through injudicious use of chemicals or through cultural practices to an extent that would permit potential pests to realize their potential and become occasional or key pests.

Agricultural and forest ecosystems vary widely in degree of stability, complexity, and the size of the area occupied. Each of these factors has important effects on pest populations and consequently affects pest-management practices. The kinds of crops, agronomic practices, changes in land-use patterns, and weather are important elements of stability. All except weather are subject to modification and therefore may be manipulated to influence pest-management practices. Variation in stability is influenced also by the duration of the host life cycle and length of the growing season.

The constantly changing, dynamic nature of agricultural ecosystems results in equally dynamic changes in pests and the problems they pose. Even the values of control procedures change in time and space with such variables as biological understanding, development of technological capabilities, changing natural conditions, and differences in economic thresholds. Perhaps in no other country has the magnitude of such changes been as great or their impact manifested so quickly and documented so throughly as in Israel.

During 1950–1965, agricultural ecosystems in Israel changed from that typical of primitive homesteads to modern mechanized farming. Although the changes have been equally apparent for large numbers of species, lepidopterous pests of maize and other gramineous crops provide excellent illustrations. Before the development of large-scale irrigation projects, *Sorghum vulgare* Person was the main summer cereal, and *Zea mays* Linnaeus was grown in small gardens and at margins of melon fields, for human consumption only. Irrigation allowed farmers to handle large areas of crops, to produce them throughout the summer, and to introduce new crops. At present, 12 species and several varieties of gramineous crops are being produced for grain, fodder, pasture, or industrial uses.

In 1930 there were two lepidopterous pests of major importance on summer cereals in Israel: *Sesamia cretica* Ledyard and *Spodoptera exigua* (Hübner). Eight other species had been recorded from cereal crops but were not considered to be of economic importance. By 1963, drastic changes in pest status had occurred. *S. cretica* had become more important with the change in cultural conditions that permitted three generations per year to be produced instead of one. The status of *S. exigua* remained unchanged, as did that of three other species present in 1930. Three species of *Chilo* could no longer be found, but *Chilo agamemnon* Bleszynski, first recorded in northwest Negev in 1959, had become the most serious pest of cereal crops. A previously unimportant species of *Chilotraea* had become so abundant and serious a pest as to preclude the growing of rice in reclaimed areas of the Hula swamps. Three other potentially important pests new to the country had been introduced and were widely distributed.

The situation in Israel is probably unique with regard to the rapidity of change of the agricultural ecosystem. However, rapid change appears likely to become an increasingly common phenomenon throughout the world in response to the demands of a burgeoning population for food, feed, and fiber. The need for food is increasing to the point that vast uncultivated areas of the tropics and subtropics must now be planted to cereal grains, pulses, and oilseed crops. Many new and serious pest problems may be expected to develop in such areas.

Some interesting changes have occurred in the past and are now taking place in the United States. They illustrate how man may control some of his pests very effectively, although unknowingly, by changing patterns of agriculture forced by economic and social changes. For example, pests of indigo relinquished that status as soon as development of the synthetic coal-tar dyes resulted in the abandonment of indigo culture in the United States. On vast acreages once devoted to cotton culture in the southern United States, the crop is not grown today. The boll weevil, once the major insect pest of the area, no longer maintains that status there.

The potential of indigenous species to become pests of new crops, varieties, or strains that are introduced into new areas or regions deserves serious consideration. The idea should be discarded that species, strains, or varieties of crops agronomically desirable in one area of the world can be transplanted to another without the hazard of furnishing the "bridge" necessary for some previously unnoticed pest species to span the distance from anonymity to the status of major pest, e.g., the Colorado potato beetle, *Leptinotarsa decemlineata* (Say), on potato. The experience of the Israelis in changing the pest status of a number of species by bringing irrigation, fertilizers, and new crops to the desert is well worth study.

UTILIZATION OF INDIGENOUS NATURAL CONTROL AGENTS

Relative scarcity rather than great abundance is characteristic of most arthropod species; but occasionally many and often a few species develop very high populations over some portions, but not all, of their range. Such species are exposed to important population-regulating factors. Data are rapidly accumulating to support two points: biotic control agents usually make up a major portion of environmental resistance to increases in insect numbers, and insects and related arthropods reach pest status most often as direct or indirect consequences of man's activities. Thus, it is important that full use be made of indigenous natural control agents.

One of the most widely known and successful examples of this approach to pest control is that with orchard pests in Nova Scotia. Control consisted chiefly in obtaining maximum benefit from natural control agents already established in the ecosystem; introduction of new species was avoided. A supplementary measure was to apply pesticides, but this was resorted to only when economic tolerance levels were threatened. Thus it was possible to take maximum advantage of environmental resistance.

Similar but somewhat less sophisticated systems for control of cotton insect and spider mite pest complexes have been operating successfully for decades in Arkansas and Louisiana in the southern United States, and in California in the western United States. Although wide differences in agricultural ecosystems exist between these areas, many elements of the problems are common. Both have a single key pest on cotton and multiple occasional and potential pest species. The key pest in the south is the boll weevil and in the west is lygus bugs. Destructive populations of the occasional and potential pest complexes develop as a result of chemical control measures invoked for control of the key species. In both areas, advantage is taken of the ability of the cotton plant to tolerate considerable amounts of apparent injury and to compensate for appreciable amounts of actual injury and still produce maximum yields. Thus, it is possible to delay applications of insecticides and often to reduce the number required during a season to the point that indigenous predators and parasites can prevent populations of the complex of potential and occasional pests from reaching economic tolerance levels. Such a system is aided considerably in the case of the boll weevil by high mortality in the overwintering population and susceptibility of larvae and pupae to the hot dry weather that often occurs during development of the first and second generations.

MAINTENANCE OF ECOSYSTEM COMPLEXITY

It is an accepted principle of ecology that complexity tends to exert stabilizing influences and simplicity tends to exert disruptive influences in the ecosystem. The tendency of populations of pests of crops grown in monocultures to oscillate widely in "boom-and-bust" cycles is well known and is in strong contrast to the characteristically more stable populations of pests associated with polycultures or natural communities. It would be highly desirable to work toward the development of complex polycultures as opposed to monocultures, from the standpoint of pest control. The interplanting of various species and varieties of crops in a complex polyculture may result in yields of total biomass equal to or even greater than that produced in monocultures. However, it remains to be seen whether systems can be devised that are of satisfactory efficiency for planting, cultivating, and harvesting such complex mixtures of crops. Here, too, the principle of economic thresholds must be observed. Much of man's success on earth stems from his success in simplifying agricultural ecosystems.

AVOIDANCE OF DISRUPTIVE ACTIONS

Since populations of insects and related arthropods most often increase to injurious levels as a result of man's activities, it is important that disruptive actions be identified and either avoided or minimized, when possible. Making use of the selective action of pesticides on predator–parasite–pest complexes is one way this may be achieved. Differential mortality may be obtained under field conditions by paying attention to methods of pesticide application, formulation, and dosage rates; timing; area of treatment; residual activity of the pesticides; habits of organisms involved; and innate physiological selectivity.

APPLICATION OF MINIMUM SELECTIVE HAZARDS

The principle of minimum selective hazard may be defined as the application of chemical control measures to pest populations in such a manner that target populations are kept just below economic-injury thresholds. By observation of this principle, the selection of resistant populations is avoided or delayed, the possibility of resurgence of treated populations is decreased, adverse effects on nontarget organisms and the amount of environmental contamination are reduced, and the cost of control is lowered.

Under the influence of ready availability of chemical insecticides of unprecedented effectiveness, the decade 1945–1955 was one in which a philosophy was developed to the highest of "eradicating" pests from the smallest (for example, one fruit tree or a small field), to the largest (major portions of 10 of the southern United States) of agricultural ecosystems by the application of broad-spectrum pesticides at maximum dosage rates. During the early years of experience with these potent toxicants, few biologists had the insight to anticipate any serious problems from their use. For the first time, weapons were available that were effective enough and cheap enough to be used widely with immediately obvious effects, and they were accepted and used enthusiastically. Little consideration was given to the rather strange phenomenon that there was no reduction in populations of most pest species with which the grower had to contend at the beginning of each new crop season, although larger and larger amounts of increasingly toxic materials were being used.

Failure of these highly effective new insecticides to bring about long-time reductions in pest populations has been elucidated by analysis of a mass of extremely detailed data on forest insects in Germany. The strategy of killing as many insects as possible whenever population density exceeds a certain level may not be the most advisable action. Insecticides used in this manner apparently elicit the sort of homeostatic mechanisms that force the pest to remain at high densities most of the time. Species that persist for long periods of geological time do so only if they are successful in evolving homeostatic mechanisms that allow them to adjust rapidly to catastrophes. Insects and related arthropods have been remarkably successful at making such adjustments. Insecticides are only the most recent of a long series of potent hazards with which insects have been able to deal successfully.

Seed-dressings especially illustrate the principle of minimum selective hazards. The rice water weevil, *Lissorhoptrus oryzophilus* Kuschel, is controlled in the United States by seed-dressings with aldrin at the rate of ¼ lb/100 lb of seed as well as by broadcast applications of the same insecticide at 2 lb/acre. Another pest of rice, the grape colaspis, *Colaspis flavida* (Say), is also controlled by seed-dressing with aldrin, but the rate has to be increased to ½ lb/100 lb of seed. Where the latter or both pests are present, approximately one fourth the amount of insecticide required to control the pests in broadcast applications is necessary; if only the rice water weevil is present, control may be obtained with about one eighth the amount.

Both of these rice pests have ecological features that also illustrate the necessity for considering the ecosystem. Seed treatment for control of rice water weevil loses its effectiveness in fields where high populations of weed hosts such as *Echinochloa* spp. are present. These weeds allow large numbers of rice water weevil larvae to develop in an untreated environment, from which they migrate to the roots of the rice plants. This emphasizes the necessity for practicing proper weed-control measures in order for a specific insect-control measure to be fully effective. The practice of seeding rice in flooded fields rather than drilling the seeds in dry soil is becoming an increasingly accepted agronomic practice. The grape colaspis is never a pest where seeding in flooded fields is practiced.

The use of systemic organophosphorus insecticides, such as phorate or disulfoton, as seed treatments for control of seedling pests of cotton provides an even more striking example of the principle of minimum selective hazards. Both materials, applied at rates as low as ¼ lb/100 lb of seed, provide satisfactory control of aphids, thrips, and spider mites in many areas of the southern United States for periods of approximately 1 month after plants emerge. The amount of toxicant applied per acre is almost infinitesimal, because seeding rates of cotton range from 10 to 15 lb/acre. There are obvious advantages of this method of application compared with that of at least two foliage applications, each consisting of 1 or 2 lb/acre of a chlorinated hydrocarbon insecticide. The seed-treatment method is rapidly supplanting foliage applications for control of aphids and thrips. Unlike broad-spectrum, relatively long-persistent chlorinated hydrocarbon insecticides applied broadcast to fields, seed treatments affect only the individuals of those species that feed on plants grown from treated seed.

It may not be valid to assume, as many have, that seed treatments do not affect populations of predators and parasites. The effects on populations of such desirable species may be quite drastic. There is little difference in destroying a population by exposure to the direct effects of contact with a toxic chemical and starving it by eliminating its prey with the chemical. For those species of predators or parasites that are sufficiently mobile to leave fields in which prey or hosts are absent or are present in numbers too few to support a population, starvation induced in this manner is probably of little consequence. However, populations of the less mobile predatory mites and spiders may be seriously affected. In any case, the over-all disruptive effect of seed treatments is far less than that of broadcast application of broad-spectrum insecticides.

The application of pesticides to restricted areas affords an opportunity for observing the minimum-selective-hazard principle. One of the most intriguing examples of this principle is described for the horn fly, *Haematobia irritans* (Linnaeus). The sexes of this pest have different resting sites on cattle. Males use the area around the ankles almost exclusively. By treating this area only, the proportion of males to females, and consequently the number of females inseminated, is drastically reduced, and control of the pest is obtained.

Most species occupy only a small portion of the territory suited to them, and, except for specific physical barriers, there is usually no obvious topographical–climatic restriction of their area of distribution. Much more attention should be directed at efforts to exclude pests from those areas that make up the vast territory suited to them but in which they do not presently occur. There is little difference in the strategy of war, whether it be man against man or man against pests. Basic strategy in human warfare is to prevent the enemy from occupying previously unoccupied areas and to deny him access to certain other requisites, for example, food. This is also prime strategy in pest control.

The history of invasion of the United States by pest species is well documented and shows that man has been generally unsuccessful in preventing pests from invading new territory. More than 100 important insect pests have succeeded in both invading and occupying parts of mainland United States since 1900. The numbers by successive 10-year periods are 24, 26, 24, 15, 7, and 8; and 4 since 1960. It is encouraging that the numbers have declined rather drastically since 1940. This is even more encouraging when the sharp decrease in numbers introduced is considered in the light of the revolution in transportation and the almost continuous involvement of the country in worldwide military operations that has occurred during this period. The decline is suggestive that the tactics employed have become considerably more effective than those used during the first four decades. However, it may be argued that the decline in numbers of successful invaders is the consequence of a decline in numbers of species that have the evolutionary resourcefulness to occupy new territory successfully after it has been invaded.

Allowing an average of almost two major pests per year to invade and occupy new territory during the last 65 years cannot be considered spectacularly successful. However, there is no way of measuring the success of efforts in this direction. The comparative ease with which most introduced species have extended their distribution once entry has been gained argues for change in both strategy and tactics. The spectacularly rapid expansion of the area of distribution of the spotted alfalfa aphid, *Therioaphis maculata* (Buckton), in mainland United States is a case in point.

Another case is the recent spread of the pink bollworm, *Pectinophora gossypiella* (Saunders), to previously uninfested areas of the western United States. Emergency measures, consisting of repetitive treatments of large areas with broad-spectrum insecticides, are under way to prevent further expansion of the distribution of this serious pest. The key word in such a course of action is "emergency." The emergency occurred many years ago when the pink bollworm first invaded this country. Similarly, the emergency posed by the cereal leaf beetle, *Oulema melanopus* (Linnaeus), occurred when it first invaded Michigan, at least prior to 1959 when farmers began treating for its control, and not in 1962 when eradication measures were invoked. The greatest weakness of current strategy to prevent establishment and spread of invading species into new territory is the lag between the time of invasion and the initiation of eradicative or containment measures.

Experience gained in the United States during the last half century shows the desirability of diverting a modest percentage of the sums currently being expended on control programs to two new areas of endeavor: research to provide the ecological and biological bases on which more effective control programs can be constructed; and more efficient survey and detection procedures that will allow control measures of truly heroic proportions to be invoked as soon as a new pest invader has been discovered, rather than after large areas of the country have been occupied.

It also seems reasonable that research of substantial proportions be undertaken on a number of important or potentially important pests that have not yet succeeded in invading mainland United States. Mere provincial interest in insect and related pests is an indefensible position for biologists to take, anywhere in the world, in view of shrinking distances and increasing contacts between various countries, which have come about with improvements in transportation. Methods on a global basis, developed from sound ecological bases, to prevent pests from extending their present distribution, should have highest priority as being the most desirable and economical of all pest-management systems. Otherwise, man faces the certain consequence of comparatively rapid spread of many important pests to the limits of their ecological adaptability. The sort of international cooperation that has developed for control of stem rusts of wheat and other cereal grains provides a model for what could be done to prevent spread of arthropod pests. Approximately 150 scientists at 85 locations in 40 countries are involved in this program.

CROP ADAPTABILITY TO ECOSYSTEMS

Time and place of crop production are commonly accepted as being determined by the suitability of climate, soils, and seasons. Consideration should also be given to the feasibility of producing a crop profitably in the face of virtually certain damaging attack by insects. It is just as important to conduct pest surveys as it is to conduct soil surveys and make studies of water availability and the climate of an area before deciding whether to grow a crop in such an environment. This may be of very great importance in less developed and developing nations as programs aimed at expanding and intensifying agriculture are initiated and carried out. Realistic pest management demands the application of this principle.

Production of sweet corn for the fresh market in many coastal plains areas of the southern United States provides an excellent example. The fresh-corn market demands that the product be free of damage by lepidopterous larvae as well as other pests. Because of high infestation levels of the corn earworm, *Heliothis zea* (Boddie), the fall armyworm, *Spodoptera frugiperda* (J. E. Smith), and the sugarcane borer, *Diatraea saccharalis* (Fabricius), in many areas bordering the Gulf of Mexico, and restrictions imposed by the necessity for producing a product free of excessive insecticide residues, corn free of damage from these pests cannot be produced economically by any currently available method. However, some progress has been made in areas where the corn earworm alone is involved, by using a combination of resistant hybrids and insecticide treatment. In such areas, varietal resistance provides the additional amount of control required for successful production of the crop. Realistic pest-management practices demand that the decision be made not to grow sweet corn for fresh market in the areas where all three species occur, on the same basis that a decision would be made not to grow the crop because of soil, water, or climatic limitations.

Need for adherence to the principle of crop adaptability to pest problems is also demonstrated by the necessity for producing seed potatoes for the United Kingdom in the north of Scotland, where aphid vectors of a virus are relatively rare and comparatively virus-free seed stocks can be produced.

The recent large increase in acreage of soybeans grown in the United States as a result of increased demands for oil and protein, and acreage controls on other crops, raises the question of adaptability of this crop to certain areas because of pest problems. An answer to the question is especially necessary in coastal areas where the crop faces the certainty of heavy insect infestations each year. In much of the northern two thirds of the United States, where soybeans traditionally have been produced, the crop is virtually free of attack by important insect pests. However, in many of the states bordering the Gulf of Mexico, growers are faced with the prospect of attempting to produce a crop profitably in the face of certain attack by a complex of more than 25 species. None of these insects is currently a key pest, but at least 10 of them are capable of assuming this status. There are imposing numbers of predators, parasites, and pathogens present. In addition, soybeans, like cotton, can tolerate appreciable amounts of apparent injury to foliage without effect on yield or quality of seed. These elements provide the bases for development of an effective, although crude, pest-management program. By taking maximum advantage of indigenous predators, parasites, and pathogens, and the capacity of the crop for tolerating considerable amounts of apparent injury without loss of yield or quality, use of selective insecticides can be restricted to applications required to control populations of pests that reach economic-injury thresholds. Meanwhile, research can be done to increase resistance, especially the tolerance component in soybean varieties, to as many pest species as possible, for example. Thus, it may be possible to produce the crop economically in such areas until more sophisticated pest-management systems can be developed. Otherwise, growers will be forced to abandon efforts to grow soybeans in many areas bordering the Gulf of Mexico, because the economics of production are such that an extensive chemical-control program cannot be sustained.

UTILIZATION OF ECONOMIC THRESHOLDS

A basic element in pest-management systems is the principle of economic thresholds. Before an attempt to discuss this principle, the term "pest" requires definition and some discussion against an ecological background. A pest is a living organism that occurs in numbers inconvenient to man. The term is without strict biological validity, because it relates only to human values of health, economics, comfort, and esthetics. Pest status is attained only when population densities that may cause significant reductions in yield or quality of crops and livestock are reached or when man's health or comfort is threatened.

An economic threshold in this context is usually defined as the level at which damage can no longer be tolerated and, therefore, the level at or before which it is desirable to initiate deliberate control activities. The definition probably should be amended to consider a more critical threshold density as that where the loss caused by a pest just equals in value the cost of available control measures. It should also take into account any known adverse side effects, which may have ecological consequences that cost more in values lost than is gained by control of the pest in question. Examples of this are the use of insecticides that reduce biotic control agents to the point where they are no longer effective and are not likely to become so again for some time, or the use of persistent insecticides that leave residues in soil which prevent the growing of succeeding crops on the treated areas because of the possibility of excessive residues occurring in the produce, e.g., soybeans following corn on soils treated with aldrin or heptachlor for control of soil insects. The amended definition takes into consideration the dynamic nature of the pest problem and allows for the establishment of relative criteria of injuriousness measured according to the general level of the economy in space and time. The determination of these thresholds is prerequisite to the development of any system of pest management for two main reasons: first, the level of pest populations below which damage is tolerable must be known, thus defining the ultimate objective of the control system; second, the level must be known above which new emergency elements of the integrated program must be applied or invoked to avert significant injury and an outbreak of the pest organism.

To obtain this information, a clear picture of the complex economic factors associated with the production of the crop of interest is vital. First, the general economic picture must be known, and what might be called the economic degrees of freedom must then be determined. In other words, the margin of profit on which the agriculturalist or forester is operating must be determined so that the amount he can afford to lose to the depredations of pests can be assessed. Second, and against this background, how much can be afforded for protection against this level of loss must be established. For example, if a grower can afford to lose *X* hundreds of dollars per acre to pests and still show a reasonable profit, he can afford up to but not more than *X* hundreds of dollars for protection of his crop by use of the control methods available to him. If his crop can be protected for less than this amount, the difference will be added to his profit. This knowledge defines the problems more clearly and sets the limits on the cost and value of the management systems that can be developed. See Chapter 18 for further discussion of the economics of pest control.

It is difficult to determine economic thresholds and levels of tolerance because of the great number of factors involved and because many of the factors are economic and not readily available to or assessable by animal and plant scientists. It is axiomatic that the threshold levels will change constantly with changing economic and environmental conditions; they themselves are dynamic, and this adds weight to the arguments in favor of multifaceted, flexible control systems.

There have been few analyses of the economics of crop production relative to pest problems, and principles have rarely been developed or limits clearly defined. Consequently, it is not unusual for more to be spent to control a pest than the value of the commodity the pest could destroy, or, even worse, for helpful insects to be destroyed at considerable cost. Moreover, the application of a pesticide to destroy pest *A* may upset balances to such an extent that new pests, *B* and *C*, are created, which in turn require still more money to control. This sort of synergism strains the boundaries of even the most liberal margin of profit and yet can only be clearly exposed by a detailed analysis of the economics of crop production in relation to pest control.

On the basis of the available fragmentary evidence, it may be concluded that economic threshold levels are almost invariably higher than expected. Too frequently, the visual threshold, the population level at which individuals of the pest species are obvious, is synonymous with the action threshold, and both are equated with the economic threshold. The action threshold is the level of pest population at which action must be taken to prevent the population from rising to the economic threshold where significant damage occurs. Ideally, it would be desirable to have control systems that are so effective and self-perpetuating that the necessity for action is avoided. Most systems, however, will probably require periodic action; thus the determination of the action and economic thresholds becomes of supreme importance.

The level of economic injury is difficult to determine. This is often assumed to be the level at which significant number of pests destroy important quantities of product, but in most cases it is a subjective determination. Damage that causes nutritive loss or adversely affects the usability or palatability of the product is far more important than damage that merely affects appearance. Esthetic values may easily be overemphasized in this regard, and modern marketing and sales promotion tend to encourage this error. The harmful effects of even quite notorious pests have proved to be remarkably difficult to document, and the very natural assumption that if a pest destroys a fruit or feeds on a tree it is causing economic injury is not always supported by the facts. It is even more difficult to determine the economic thresholds of most pests on most crops.

PREDICTION OF POPULATION TRENDS

Once the economic status of the pests occurring in an ecosystem has been determined, studies on the biology and ecology of the pests must be developed. These have two purposes: prediction and manipulation. The main value of being able to predict future trends in the population levels of pests is that it enables the intelligent application of control measures to prevent rises above the economic-injury level. Most pest-management programs will be complex interwoven systems with a number of major components. Pest populations will certainly not be eliminated in these programs but, rather, will fluctuate at low levels generally acceptable to man. From time to time these fluctuations will approach the economic-injury level. If reaction to such fluctuation is in the application of vigorous extra measures, the system may be permanently disrupted. Therefore, means for predicting future population trends with accuracy and confidence must be developed, so that new components may be added to the system only when required to damp potential outbreaks; in this regard components must be selected that have a minimal disruptive influence on the system as a whole. When danger is past, these components should be dropped from the system until they are again required. This kind of prediction is particularly important after a satisfactory system has been developed—in the operational phase of the study.

Manipulation is also basic to the establishment of such systems. The factors is an agricultural ecosystem that affect pest numbers, or that have the potential to do so, must be determined, and those with the greatest utility must be selected and so manipulated that their regulative effect in the ecosystem as a whole will be maximized. When this has been done and the most value is being obtained from natural factors of the ecosystem, it may be found that the pest populations of interest are reduced to tolerable levels without further action. If not, means of adding new components to the ecosystem may be considered; these components will complement those already present to produce the required levels of pest abundance. In the rational and organized development of pest-management systems, this step-wise progression in research, from determining important environmental factors, through manipulation to maximize their effectiveness, to the addition of supplemental components, is essential.

MAINTENANCE OF SUBECONOMIC POPULATIONS

Full development of integrated control systems requires the continued existence of pest populations in the ecosystem. This is the most difficult aspect of the integrated-control concept for many to grasp. Low pest densities may need to be conserved to maintain the pest-management system and to provide the necessary elements of permanence and dynamism. If the pest is completely eliminated, many of the regulating components of the ecosystem also may disappear. Pest-management systems often cannot operate without the presence of pests in subeconomic populations. In the absence of these critical components, the pest-management system can no longer function as a self-generating system if the pest reinvades the ecosystem, as it is almost certain to do. The subeconomic pest population and its associated natural controls need not always be in the crop-growing area. With highly agile populations, they often are not but may be in associated uncultivated areas, hedgerows, weeds, and cover crops.

Pest populations can be tolerated at levels that have no effect on crop yield and quality. The mere presence of an insect pest is not a threat of economic damage, and pests do not have to be eradicated to be controlled.

SUCCESSFUL UNILATERAL PEST-MANAGEMENT SYSTEMS

Systems of pest management based on the single-factor approach are proving to be less and less reliable. Nevertheless, there are examples in many unilateral systems that have worked successfully for decades and show no evidence of declining in effectiveness. All such examples show evidence of having been developed from intimate knowledge of biological and ecological principles. The various approaches have been based on manipulation of ecological niches, with the objective of disrupting ecological requirements or of augmenting ecological intolerances.

Biotic requirements such as food, breeding sites, and shelter are vulnerable requisites for pest species. Denying access to one or more of these is an enlightened ecological approach that has been demonstrated to be effective. Two of the most successful demonstrations of the approach have been the complete control of the grape phylloxera in the vineyards of Europe by use of American grape rootstocks on which the insect cannot complete its life cycle, and control of the Hessian fly, *Mayetiola destructor* (Say), in many areas by use of varietal resistance in wheat.

Satisfactory control of the southern cattle tick, *Boophilus microplus*, (Canestrini), on cattle in Queensland has been obtained by denying the pests access to their hosts for periods long enough to break the life cycle. This technique of pasture "spelling" was based on data showing that ticks detaching from their hosts during mid-April to mid-July produced few or no progeny. The reasoning that formed the basis for the approach was that if cattle were moved to an uninfested or sufficiently lightly infested pasture in May and kept there until larvae in the alternate pasture died, the life cycle would be broken, and that moving cattle at 4-month intervals thereafter would prevent any small residual infestations from increasing to troublesome levels.

An example of excluding a pest from its host by temporal separation of the two is provided by a system used for control of the pickleworm, *Diaphania nitidalis* (Stoll). Cucumber can be profitably grown as either a spring or fall crop in much of the temporary range of the pickleworm. The method takes advantage of this fact plus detailed knowledge of the biology of the pest, which is endemic in areas bordering the Gulf of Mexico. The insect disperses northward each year, extending its range with each generation until it is eliminated from its temporary range by cold weather. It is successfully controlled by producing pickling cucumbers as a spring crop, before the annual invasion of the pest, thus avoiding infestations.

Control of a pest complex has been obtained in Egypt simply by changing the planting date of maize, *Zea mays* Linnaeus. Prior to development of this method, maize was planted throughout April to September, and the stem borer complex–*Ostrinia nubilalis* (Hübner), *Sesamia cretica* Ledyard, and *Chilo agamemnon* Bleszynski–caused such severe damage to the crop that growers had to resort to extensive application of insecticides for control. By confining the planting dates to mid-May through mid-June, excellent control of all species has been obtained, and yields have increased substantially.

FROM CHEMICAL TO INTEGRATED CONTROL PROGRAMS

It is too often assumed that once a unilateral pest-control program based on multiple applications of broad-spectrum insecticides has been adopted it is impossible to revert to an integrated control program. It may be true that when such a program fails and a multifaceted pest-management system has to be substituted there is likely to be a difficult period of transition. During this time the grower may have to be willing to sustain a considerable amount of damage until corrective measures have operated long enough to allow the necessary factors to become effective in the ecosystem. Such transitions can be made successfully, as illustrated by the well-known work on orchard pests in Nova Scotia, the spotted alfalfa aphid investigations in California, and the less publicized but equally impressive studies on cotton pests in Peru.

In Peru, cottongrowers of the Canete Valley were forced to abandon the single-component system of complete reliance on application of broad-spectrum insecticides because of the development of insect resistance to insecticides, resurgence of treated populations, and elevation to pest status of several species of minor or no previous importance. Reasonably complete records for more than 30 years are available to chronicle the history of pest control on cotton in this narrow, isolated valley.

Prior to 1939, only six species were considered to be pests in Peru, and they were controlled by the use of arsenical and nicotine insecticides and hand picking of insects and infested fruit. These methods allowed a high degree of natural control to be exerted by indigenous predators and parasites. By the 1949–1950 crop year, virtually complete reliance had been placed on the applications of large amounts of broad-spectrum insecticides, such as DDT, benzene hexachloride, and toxaphene, or mixtures of these with sulfur. The most important pests were the tobacco budworm, *Heliothis virescens* (Fabricius), and the cotton aphid.

Results of the new program were good during the first 5 years. Good control of the pests was obtained with consequent increases in yields. However, after the third year it was observed that the insecticides were losing their former effectiveness. New insecticides, such as aldrin, dieldrin, and endrin, and various mixtures of these, were used, and intervals between applications were reduced. This increased costs and failed to improve the situation in many cases, and growers changed to the use of organophosphorus insecticides, such as parathion and mixtures of ethyl and methyl parathion.

After these insecticides were used for a few years, six new and important pests appeared. The situation deteriorated rapidly as several of the pests developed levels of insecticide resistance bordering on immunity, and it culminated in 1956 with the worst crop in the history of cotton production in the Canete Valley. Nearly 50% of the crop was lost after an average of 12 applications of insecticides failed to give adequate control of the pest complex.

Based on experience to this point, both growers and experiment station personnel concluded that it was not possible to continue profitable cotton production based on chemical pest-control programs, and in 1957 a pest-management system was adopted that contained elements of chemical, cultural, biological, and regulatory methods. The system was based on the following key points: (1) reduction in the planting of ratoon cotton (crop produced from plants cut back and allowed to produce new growth from the crown) to less than 25% of the total area; (2) preparation of soil without irrigation to obtain increased destruction of *Heliothis virescens* pupae; (3) repopulation of predators and parasites by introduction from other valleys or foreign countries, or by artificial rearing for release; (4) establishment of mandatory planting and crop-residue destruction dates; (5) adoption of recommended irrigation schedules; (6) adherence to the various cultural methods considered to be sound agronomic practice for production of uniformly early crops; and (7) limiting insecticides used to arsenical and botanical materials only, except under very special circumstances, which allowed application of systemic insecticides at dosage rates one fourth to one half of that recommended by the manufacturer to control *Aphis gossypii*.

Seven years after the initiation of this program, apparently the insect problems have been solved and the natural balance between detrimental and useful insects has been restored. More than 30 species of predators and parasites are currently playing an important role in control of the pest complex. Finally, yields have increased by about one third compared with the 1950–1955 period, when control was based on the exclusive use of insecticides.

DEVELOPMENT OF INTEGRATED SYSTEMS

The purpose of this section is to describe the information required for the development of integrated pest-management systems, to explain how this information may be obtained, and, finally, to discuss the ways in which it can be used to design such systems.

The problem of building an integrated control program by organized, deliberate effort is essentially one of systems design. The problem of obtaining the information that will furnish the building blocks for this design is largely one of analyzing various systems. Researchers have taken a remarkably unsophisticated approach to these problems, an approach highly colored by traditional training and experience as applied biologists and largely ignoring the development of tools with tremendous potential value by theoretical biologists and biomathematicians. This fault is not entirely one-sided. If traditionalists could learn to become more mathematical and theoretical, and if the new school could become more biological, an effective collaboration could develop, with potentially great revolutionary consequences.

Living organisms are multicomponent open-chain systems that combine remarkable stability with great sensitivity to change. Such systems are difficult to describe, let alone analyze, and yet they must be analyzed if pest-management systems are to have a rational and firm basis in fact and understanding. Without mathematics it is hardly possible to think about entities with more than two or three variables, and yet there may be thousands of variables in a biological system. The problem is to select those that must be studied and to determine how few of them are adquate to explain the system's behavior.

Many ecologists object to this approach, protesting that ecosystems and population processes are too compex to allow hope for detailed and precise analyses. However, there is little in physics and chemistry to rival the striking success and broad sweep of genetic theory. It stands as a rebuttal to those who maintain that biological phenomena are too complex to yield to theoretical analysis. Applied ecologists must either equip themselves for such analyses or interest the theoreticians in the problems. The importance of agricultural and forest industries to over-all economic well-being fully justifies this attention, but the ultimate applied nature of the problem may repel the theoretician.

Systems theory is a special kind of formal approach to biological problems. It is based on the modern development of information theory, game theory, operations research or cybernetics, and computer development. Everyone is familiar with the notion that the whole is greater than the sum of its parts. The concept of systems and systems theory is closely tied to this idea. In other words, a system of interacting entities has certain properties that are not obvious properties of the component entities themselves. The primary problem in studying such entities has been to predict the properties of the whole from the rules of interaction between its components. The type of question posed is how the properties of a biological system are the result of their being a complex array of interacting elements. Homeostatic problems have been the core of much of the biological work on systems theories; these also lie at the heart of understanding population phenomena.

In formal treatments of organizations, such as an ecosystem, much use is made of a simple conceptual device called the black box. This name is used for an object about which the following is known: it can sense a certain repertory of inputs, generates a certain repertory of outputs, and associates outputs to inputs according to a certain repertory of admissible laws. The black box can represent anything desired, for example, a particular behavioral feature of an organism, such as predation. If the jargon is stripped from the above statement, what emerges is that predacious behavior (output) is a response to certain environmental factors, both internal and external (input), and that its expression falls within certain limits (the laws).

The black-box concept had its origin in systems engineering. In the design of a complicated system, it is difficult to consider all parts at the same time. Therefore, an attempt is made to separate the function of the entire system into independent component functions, each of which is represented by a black box. When this has been done, separate consideration can be given to the problem of how to replace each black box by some real object or activity that will yield the desired function. This describes exactly the problem of designing integrated control systems and presents a new, simple, and refreshing view of a way in which it can be approached.

Detailed information on many components of the ecosystems in which pests operate is essential to the design of integrated control systems, and obtaining this information can be viewed not only as a problem in ecology, but also as one in systems analysis. The black-box concept is also useful in the analysis of systems. In studying any particular function of a living system, it is not necessary to analyze the nature of every single factor involved; some factors can be replaced by black boxes, and the analytical effort can be concentrated on one or a few factors. In either context, the black box is merely a symbolic representation of functions associated with a given object or class of events. The progress of both systems design and analysis can be symbolically represented by a progressive replacement of black boxes by white boxes. A white box is a system of known components put together in a known fashion to produce a given result. The replacement of every black box by a white box is a possible solution of the task. Although the contents, or properties, of a black box are not known, the usual purpose of isolating it as a unit is to discover or hypothesize what they are. To do this, the black box is tentatively replaced in the over-all system by a white box whose contents are known and have deliberately been chosen so that the input–output relations of the white box will be the same as those observed for the black box it represents. The white box, therefore, is a model, hypothesis, or explanation of how the black box works.

There are several principal mathematical theories of systems, such as cybernetics, game theory, and communication theory. Cybernetic analysis has merits that are particularly revelant to a consideration of integrated control systems; one is the way in which it permits the partition of systems design and analysis into clearly separated components, or black boxes, for specific study. Another is that it recognizes the general importance of feedback, or the guidance of control actions through sensing or monitoring their results.

Game theory deals with situations of conflict between two or more objects or procedures. Input and output repertories difficult to specify in natural situations are replaced by a repertory of well-defined possible moves. The mathematically difficult problem is that of generating the strategy, or the law that prescribes a move for every possible situation. The basic rule is to design a strategy such that the expected gain is maximized (for example, crop yield) if the opponent (pest) follows what for him is an optional strategy (perhaps reproduction, feeding, or movement). This is what is being attempted in designing integrated control systems, but the methods are empirical and imperfect; they are based more on natural history and descriptive ecology than on theoretical biology. The advantages of placing efforts on a sounder basis of theory and of taking advantage of the techniques of systems analysis and design are obvious.

The claim will be made that the approach is too sophisticated for an essentially field-oriented applied problem. If the theories of systems analysis and design were adopted by applied ecologists concerned with pest control, they could be adapted to meet specific aims. This would require basic research in biomathematics and theoretical biology, but once adapted as a routine tool to be used for integrated control systems, shortcuts and simplifications would undoubtedly follow that would place the tool within the reach and comprehension of all.

Systems analysis and design have already been used to advantage in problems of fisheries and wildlife management. Moreover, some entomologists are using systems analysis in studying population processes, such as predation and attack, that are of direct and significant interest to those using predators or parasites as a biological control component of pest-management systems. Other ecologists are attempting, on a purely logical basis, to study problems of pest control by separating the biological systems of concern into their component parts, or black boxes. They could benefit greatly by adopting not only the logical basis of systems analysis and design but also the methods of mathematics that are already available for this approach.

There are many advantages in opening the doors of a narrow, problem-oriented research world to the strengths that have been developed elsewhere. Some of these are: Approaching research on incredibly complicated population phenomena through systems theory would bring order, organization, and coherence to efforts expended; the information necessary for pest-management systems could be more clearly defined, and data for use in designing such systems could be processed more efficiently; encouraging the analysis of individual components would lead to simplicity rather than what has been called a "sea of confused complexity"; advantage could be taken of the principles of systems analysis and design already developed, but largely unknown to applied ecologists; advantage could also be taken of the mathematics for handling systems that are now available and being actively developed; and, finally, computer programs for systems analysis and design are already available that could be adapted to the problems.

With this background, and bearing the theories of systems analysis and design in mind, consideration is given to the information required for the perfection of pest-management systems. This is a matter with two distinct aspects: the analyses of the ecosystems with pest problems that will provide the needed information; and the ways in which this information will be used to design control systems.

The major difficulty in determining which factors in an ecosystem regulate pest populations, or have the potential to do so, is that of gathering the complex data required and of analyzing them to extract the required information. One particularly useful tool, the life table, is now available. This can be a highly sophisticated research device or a relatively simple tool, depending on the orientation and interests of the research worker using it. In the development of pest-management systems, a simple form of the life table will probably meet the needs satisfactorily. This can be thought of as a profit and loss analysis of the dynamics of the pests, with the determination of the causes of major losses, or mortality. Analyzing the dynamics of any organism to determine the most important factors can be done simply by determining the life-cycle stage most responsible for increasing or decreasing numbers and then pinpointing the causative factors within this stage. Percentage of mortality is not good measure of the importance of a factor affecting the dynamics of a pest; variation is the important attribute of mortality, and low but variable mortality may have considerably more importance than high but consistent mortality.

It is difficult to believe that satisfactory control systems can be developed rationally without an understanding of the dynamics of the pests of major concern. As stated previously, such a foundation permits maximum value to be taken of factors already operating in the ecosystem to cause mortality to the pest, and then, but only then, to add new factors as required.

Successful pest-management systems will probably have many components, and research in many areas will be pertinent to their evolution, with varying degrees of directness: toxicology, plant resistance, genetics, and almost any field of biology that can be conceived. Certain related research fields, however, are of paramount importance to the perfection of management systems: (1) bioclimatology, because without an understanding of the ways in which weather affects population levels and trends, prediction is impossible; (2) toxicology and studies of the effects of different kinds of pesticides on both harmful and beneficial insect species, better understanding of the mechanisms of resistance to insecticides, and ways of managing populations to avoid or delay the development of resistance, because, without this information, pesticides that will affect the pests in desirable ways and yet have minimal effects on beneficial species cannot be incorporated; (3) studies of the principles of parasitism and predation, because without knowledge of these principles it will not be possible to put the selection of beneficial species for inclusion in the system on a rational basis; and (4) studies of the effects of host-plant resistance on populations of predators and parasites and the effectiveness of pesticides.

The need for knowledge has so far been explained on the basis of economic relations between pest populations and crop production, and the requirement for information on the dynamics of pest populations so that existing factors can be manipulated and thus have their maximum influence, and so that new factors can be introduced to the system to complete the process of regulating densities at acceptable and satisfactory levels. One of the most difficult problems must now be considered: selecting the components to be added to the system and ensuring harmony among those selected with the entire ecosystem background. It must be ensured that the components of the management program are fully compatible, perhaps even synergistic, and do not disrupt or interfere with one another. This is essentially a problem in optimization, and much can be learned of this from advances made in other fields, such as fisheries and wildlife management. The only major differences between these fields and entomology is that they are concerned with maximizing animal populations as the harvestable commodity, whereas entomologists are interested in maximizing livestock and crop production at the expense of pest populations. The principles are the same.

When thought about in terms of devising systems for the entire array of pests in a crop ecosystem, enormous complexity is being added to the problem of harmonizing the components of the system. Actually, any specialist in computer technology can help find the solution of this problem of complex, multivariate systems, but entomologists have not yet fully realized this great potential. Computer games such as simple numerical routines that mimic the effects of environmental factors to simulate pest populations without control and with a variety of insecticidal applications, or dynamic programming, a computer device involving multistage decision processes, could be used to determine optimal control strategies. These exercises in the game-theory aspect of systems design have value for synthesizing real management programs. All that is needed to realize this value are data collected from actual field situations. These can be supplied through research.

A number of theoretical considerations indicate that the integration of two or more techniques into a unified pest-management system should provide more effective and less costly control of most pests, and pest complexes, than the use of any one method alone. It has been generally accepted, for example, that the use of conventional insecticides is highly efficient in terms of numbers of insects destroyed at high population densities but is relatively inefficient when populations are low. Conversely, release of sterile insects is inefficient at high-population and efficient at low-population densities. Integration of the two methods should provide a system of pest management much more efficient than either used alone, especially if the sterile-insect release component should provide a means of reversing the law of diminishing returns, as has been postulated, that operates in all unilateral systems.

In an area occupied by a pest, the cost of attaining any given level of control by an agent that actively seeks the pest will be proportional to the total population and largely independent of area, while the cost of a passive agent that does not seek the pest will be proportional to area and independent of the population. For example, if an insecticide gives 90% control of a pest population, it would be expected to do so whether the pest density was 1 or

100 per unit area, and the cost would be the same. The same result would be expected from the application of pathogens, field treatment with chemosterilants, or the use of attractants or traps.

The cost of rearing the number of parasites, predators, or sterile males required to attain any given reduction in a pest population will be proportional to the size of the population. Only the cost of distribution, generally a small part of the total, will be partially dependent on area. Therefore, passive agents are likely to be cheapest when pest densities are high, and active agents should be less expensive at low levels. Suppose, for example, that pest density averages 10,000 insects per acre, with a rate of increase of five times per generation. Suppose that the economic threshold is 2,000, that an insecticide application giving 80% control costs $8 per acre, and that one application per generation will be required to hold the population as the economic threshold. Now suppose that parasites cost $2 per thousand and that each parasite destroys one host if enough are released to give 80% control. That would be 8,000 parasites costing $16 per generation, or twice the cost of insecticides. However, suppose that an insecticide was first used to reduce the population to 2,000, then this was followed in the same generation with the release of parasites to reduce it another 80% to 400 hosts. The parasites would cost $1.60. In the first generation, then, the cost would be $8 + $1.60 = $9.60, but the continuing cost of holding the population at the 400 level would be only $1.60 per generation, as against $8 for insecticides alone or $16 for parasites alone at the higher mean population.

Still another theoretical advantage of multifaceted pest-management systems results from the law of diminishing returns, which applies to all methods. If the intensity of any agent is increased to achieve a higher kill, there is a point beyond which the dose or the work required will increase exponentially as control increases, and the efficiency per unit of cost will decrease proportionately. For instance, the quantity of an insecticide required to kill insects increases exponentially as the percentage of control increases. In one experiment in which schradan was applied to the corn planthopper, *Peregrinus maidis* (Ashmead), a concentration of 1.5 mg/ml of solution killed 50% but 6.25 mg/ml, or more than 3.5 times as much material, was required to kill 100%.

The same rule holds for the release of sterile males. Theoretical calculations, well supported by experimental data, show that to increase control from 30 to 60 and 90% would require an increase of 3.57 times and 21.42 times in the numbers of sterile males released instead of 2 and 3 times, as might be expected.

Many laboratory and some field experiments show that the exponential rule also applies to the release of parasites or predators. Thus, in Canada, pupae of the European pine sawfly *Neodiprion sertifer* (Geoffroy), were placed on a lawn in the open, and varying numbers of the parasite *Dahlbominus fuscipennis* (Zetterstedt) were released. The control was 11.3% with 25 parasites, 19.0% with 50, and 40.1% with 200. When workers in Poland released *Trichogramma* against the plum moth in orchards, the control was 81.5% with 2,000 parasites per tree but only 91.3% with 20,000.

The rule of exponential increase in cost means that it will often be cheaper to use two methods in sequence instead of one alone. Whether or not there will be a saving, and the amount, will partly depend on the cost of the methods used. Obviously, two methods will not be cheaper than one if one of them is very expensive. The saving will also be influenced by the proportions of the total cost that can be charged to materials or application. For example, doubling the degree of control might take 4 times as much insecticide, but it would cost no more to apply it. Application of computer technology can greatly improve thc efficiency of experiments designed to study problems of this sort under actual field conditions.

In summation, there are three clearly defined primary research needs for the development of integrated pest-management systems. First, research is needed on the economics of crop production relative to pests and pest control, and on economic injury levels relative to individual pest species. This is to clarify the importance of the pest problem so that the required intensity of population regulation can be defined. Second, analysis of the dynamics of the pest populations relative to mortality factors already operating in the ecosystem is necessary to provide the information required to predict future population trends and to enable the most effective factors to be intensified through modifications or manipulations of the environment. Third, supplemental methods of control available for use in the particular ecosystem of interest must be evaluated to determine which are most suitable for the needs, and to harmonize these into the existing system to produce optimal crop production relative to pest infestations.

The pests against which the successful management system is developed, as well as other components of the ecosystem, are dynamic. Therefore, the management system itself must be dynamic and have great flexibility. This means that when an integrated program has been developed and applied, it must not be expected that the job has been finished; continual changes in its details will be needed to adapt it to the dynamic ecosystem. New components must be added and former ones removed as their usefulness passes. To maintain this flexibility, there will have to be frequent monitoring of the major components of the ecosystem. This is not primarily a research activity, but rather maintenance most suitably accomplished by extension personnel or by sophisticated growers themselves. It will be based on the research date of the developmental phase but will not itself require great precision and detail. If background data are adequate, shortcuts to sampling and analysis can be taken. This operational phase might require as little as merely making periodic visual assessments of a control system or as much as regular sampling or trapping to follow population changes.

The research worker required to meet the needs of this rapidly evolving philosophy of pest control will be an entomologist trained for research on pest-management systems—one who is first and foremost a biologist in the broadest sense, one with a good background in traditional entomological principles, but also fully familiar with behavior, toxicology, population ecology and genetics, resource management and the techniques on which it rests, systems analysis, economics, biomathematics, and computer programming. Above all, he must have a personal outlook highly colored by flexibility, imagination, and receptiveness to new ideas. Such a person would indeed be a paragon, and the only practical way in which this array of talents can be brought to bear on the complex problems of pest control is by the formation of teams. Cooperative, interdisciplinary research and the team approach are becoming as essential to the evolution of pest-management systems as is the meeting of these research needs.

BIBLIOGRAPHY

Bartlett, B. R. 1956. Natural predators. Can selected insecticides help to preserve biotic control? Agr. Chem. 11(2):42–44, 107–109.

Beirne, B. P. 1967. Pest management. Leonard Hill, London. 123 pp.

Boza-Barducci, T. 1965. Experiments about the utilization of the biological control methods and the integrated control methods of cotton pests in Canete Valley—Peru (S.A.). Unpublished paper presented at the 24th Meeting, Cotton Committee Production Investigation, Washington, D.C., May 18–19.

Chant, D. A. 1964. Strategy and tactics of insect control. Can. Entomol. 16:182–201.

Chant, D. A. 1966. Integrated control systems, pp. 193–218. *In* Scientific aspects of pest control. NAS-NRC Publ. 1402, National Academy of Sciences–National Research Council, Washington, D.C.

Food and Agriculture Organization of the United Nations. 1966. Symposium on integrated pest control, October 11–15, 1965, Rome. (D. F. Waterhouse, Chairman). Proceedings, Part 1, 91 pp.; Part 2, 186 pp.; Part 3, 129 pp.

Franz, J. M. 1961. Biological control of pest insects in Europe. Annu. Rev. Entomol. 6:183–200.

Geier, P. W. 1964. Population ecology and pest control. J. Aust. Inst. Agr. Sci. 30:23–28.

Geier, P. W. 1966. Management of insect pests. Annu. Rev. Entomol. 11:471–490.

Geier, P. W., and L. R. Clark. 1960. An ecological approach to pest control. Proc. 8th Tech. Meet. Int. Union Conserv. Nature and Natur. Res., Warsaw, pp. 10–18.

Glen, R. 1954. Factors that affect insect abundance. J. Econ. Entomol. 47:398–405.

Harley, K. L. S., and P. R. Wilkinson. 1965. A comparison of cattle tick control by "conventional" acaricidal treatment, planned dipping, and pasture spelling. Aust. J. Agr. Res. 15:841–853.

Hikichi, A. 1964. The status of apple leaf rollers in Norfolk County, Ontario. Proc. Entomol. Soc. Ontario. 94:38–40.

Knipling, E. F. 1966. Some basic principles in insect population suppression. Bull. Entomol. Soc. Amer. 12:7–15.

Knipling, E. F., and J. U. McGuire, Jr. 1966. Population models to test theoretical effects of sex attractants used for insect control. U.S. Dep. Agr. Inform. Bull. 308. 20 pp.

Lawson, F. 1958. Some features of the relation of insects to their ecosystems. Ecology 39:515–521.

Morris, R. F. 1957. The interpretation of mortality data in studies on population dynamics. Can. Entomol. 89:49–69.

Morris, R. F., and C. A. Miller. 1954. The development of life tables for the spruce budworm. Can. J. Zool. 32:283–301.

Newsom, L. D. 1962. The boll weevil problem in relation to other cotton insects, pp. 83–91. *In* Proceedings boll weevil research symposium, State College, Mississippi, March 1962.

Paradis, R. O. 1956. Factors in the recent importance of the red-banded leaf roller, *Argyrotaenia velutinana* (Walker) (Lepidoptera:Tortricidae) in Quebec apple orchards. 38th Rep. Quebec Soc. Prot. Plants. pp. 45–48.

Pickett, A. D., N. A. Patterson, H. T. Stultz, and F. T. Lord. 1946. The influence of spray programs on the fauna of apple orchards in Nova Scotia: I. An appraisal of the problem and a method of approach. Sci. Agr. 26:590–600.

Pickett, A. D., and A. W. McPhee. 1965. Twenty years' experience with integrated control programmes in Nova Scotia apple and pear orchards. Proc. 12th Int. Congr. Entomol., p. 597.

President's Science Advisory Committee. 1965. Restoring the quality of our environment; Report of the environmental pollution panel. The White House. U.S. Government Printing Office. 317 pp.

Quastler, H. 1965. General principles of systems analysis, pp. 313–333. *In* Theoretical and mathematical biology. Blaisdell Publishing Co., New York.

Rivnay, E. 1965. Present status of lepidopterous pests of maize and other gramineous crops in Israel. FAO Plant Prot. Bull. 11:1–3.

Smith, R. F. 1962. Principles of integrated pest control. Proc. N. Cent. Br. Entomol. Soc. Amer. Vol. 17.

Stern, V. M., R. F. Smith, R. van den Bosch, and K. S. Hagen. 1959. The integration of chemical and biological control of the spotted alfalfa aphid. I. The integrated control concept. Hilgardia 29:81–101.

Thompson, H. V. 1960. Economic ornithology. Ann. Appl. Biol. 48:405–408.

Turnbull, A. L., and D. A. Chant. 1961. The practice and theory of biological control of insects in Canada. Can. J. Zool. 39:697–753.

Ullyett, E. C. 1951. Insects, man and the environment. J. Econ. Entomol. 44:459–464.

Watt, K. E. F. 1961. Use of a computor to evaluate alternative insecticidal programs. Science 133, March 10, 1961.

Watt, K. E. F. 1963*a*. Mathematical population models for five agricultural crop pests, pp. 83–91. *In* E. J. LeRoux (ed.). Population dynamics of agricultural and forest insect pests. Entomol. Soc. Can. Mem. 32.

Watt, K. E. F. 1963*b*. Dynamic programming, Look ahead programming, and the strategy of insect pest control. Can. Entomol. 95:525–536.

Wene, G. P., R. A. Blanchard, and E. V. Walter. 1952. Relation of corn earworm (Heliothis Armigera) resistance in sweet corn to efficiency of insecticide sprays. J. Econ. Entomol. 45:931–933.

AUTHOR INDEX

KEY-WORD TITLE INDEX